ELEMENTARY
INTEGRAL CALCULUS

ELEMENTARY
INTEGRAL CALCULUS

By

G. LEWINGDON PARSONS, M.A.

SOMETIME SCHOLAR OF CHRIST CHURCH, OXON
MATHEMATICAL MASTER AT MERCHANT TAYLORS' SCHOOL

CAMBRIDGE
AT THE UNIVERSITY PRESS
1926

CAMBRIDGE
UNIVERSITY PRESS

University Printing House, Cambridge CB2 8BS, United Kingdom

Cambridge University Press is part of the University of Cambridge.

It furthers the University's mission by disseminating knowledge in the pursuit of education, learning and research at the highest international levels of excellence.

www.cambridge.org
Information on this title: www.cambridge.org/9781316612644

First published 1926
First paperback edition 2016

A catalogue record for this publication is available from the British Library

ISBN 978-1-316-61264-4 Paperback

PREFACE

THE object of this little work is to provide an introduction to the subject of Integral Calculus for mathematical and scientific students. The plan of the book is largely based on the Syllabus for the Higher Certificate Examination and it owes its origin to a set of lectures given by the author in preparation for this examination. The author ventures to hope that, while reducing the ground to be covered, he has not left out anything which is fundamentally necessary to the opening stages of the subject. A previous knowledge of the differentiation of powers, products, sines, cosines, etc. is however assumed.

No attempt has been made to treat the subject from an absolutely mathematical standpoint. The author fails to see the value of discussing the validity of a process until the nature of the process and its usual applications are thoroughly grasped. In fact, the work is not by any means intended as the final reading for a prospective mathematical scholar.

It is the author's opinion that, in many books, sufficient stress is not laid on what he has ventured to call the Fundamental Theorem. Historically the discovery of this theorem by Newton and Leibnitz was the turning point of the whole theory. He has therefore reversed the conventional order of treatment of the subject, and introduced the notion of summation only after the student has learnt how to integrate. It is hoped that this will lead to a deeper understanding of the Fundamental Theorem, and hence to a more skilful use of it.

In view of the recent regulations permitting the Preliminary Examination in Natural Science at Oxford to be taken from school, a chapter on Differential Equations has been added, making the book a complete preparation for the Integral required. The addition of questions on rotational dynamics to the Mechanics Syllabus of the Higher Certificate Examination seems to indicate a move in this direction, whilst from the point of view of general education the popularity of wireless renders some mathematical knowledge on the subjects of "damping" and "resonance" highly desirable.

A short historical survey has been added, for the author believes that this is a subject which might, with advantage, be taught to a greater extent than is done to-day.

His acknowledgements are due to Mr T. W. Chaundy, M.A., Tutor of Christ Church, Oxford, who has, in the press of much more important work, been good enough to read the MS. and has made many valuable suggestions. The extent to which the author is indebted to many standard works will be apparent to any mathematician who reads this volume. He desires to mention in particular the works of Williamson, Gibson and Edwards.

Questions from the Higher Certificate Examination are included by kind permission of the Oxford and Cambridge Schools Examination Board, and are marked in the text with an asterisk.

In conclusion the author wishes to record his indebtedness to the readers of the Cambridge University Press, to whose careful and excellent work the elimination of many errors is due.

G. L. P.

November, 1925.

CONTENTS

CHAPTER V

TRIGONOMETRICAL INTEGRALS AND FORMULÆ OF REDUCTION

CHAPTER VI

THE FUNDAMENTAL "INVERSION THEOREM." APPLICATION OF CALCULUS TO FINDING OF AREAS

CHAPTER VII

FURTHER APPLICATIONS TO GEOMETRY: SECTORIAL AREAS, VOLUMES OF REVOLUTION, FINDING THE LENGTH OF A CURVE

CHAPTER VIII

APPLICATION TO PROBLEM OF FINDING CENTRES OF GRAVITY

CHAPTER IX

FURTHER APPLICATIONS TO MECHANICS. RIGID DYNAMICS

CHAPTER X

DIFFERENTIAL EQUATIONS. A FEW TYPES

HISTORICAL SKETCH

THE arrangement of the subject-matter of Integral Calculus adopted in the text is almost exactly opposite to the order of the historical development of the subject. A little reflection will convince the student that this is the case in most branches of science. Attention is first drawn to some particular problem, insoluble by known methods. A method is discovered for this particular problem, and is gradually proved to be true for a large class of allied problems. Finally a series of rules governing the use of the method is drawn up. In teaching the subject, of course, the rules have to be learnt first and the application follows. This is exactly the history of the Integral Calculus.

It was quite inevitable that the attention of mathematicians should be drawn, sooner or later, to the problem of summation. As early as the third century B.C. we find the great geometer and physicist Archimedes determining the area bounded by a parabola, by splitting it up into rectangles and finding the sum of these rectangles. He arrived at the correct result, which shows that he knew the result, which we should express in the form

$$\int_0^1 x^3 \, dx = \left[\frac{x^4}{4} \right]_0^1,$$

though, of course, he had not this notation. Very little advance was made for many years. The Hindu and Arab mathematicians who succeeded the Greek school concentrated their attention mainly on algebra and trigonometry and paid no great attention to questions of area and so forth. They were, however, responsible for the formulæ

$$\sum_1^n r^2 = \frac{n(n+1)(2n+1)}{6},$$

$$\sum_1^n r^3 = \frac{n^2(n+1)^2}{4},$$

which were of course used by later writers in performing summations.

But it is not till the time of Johann Kepler and Bonaventura Cavalieri that we find any definite traces of a general attempt to attack the problem of summation. In 1609 Kepler published his work " On the motion of the planet Mars " and it is clear that he was in possession of some method for finding the areas of focal sectors of an ellipse, whilst a few years later in another work he regards the volume of a cask as composed of thin circular discs, just as we should. Cavalieri, who was a Jesuit, published in 1635 the "Method of Indivisibles." This method, though his statement of it was not mathematically rigid, was in effect the method of elements that we use to-day. He stated, for instance, that a line is composed of a number of points without magnitude, a plane area of a number of lines without breadth, and a volume of a number of planes without thickness. He also arrived at the result

$$\int_0^a \frac{b^n x^n}{a^n}\, dx = \frac{ab^n}{n+1},$$

which is equivalent to the rule for integrating x^n.

The method of indivisibles found many exponents, and in the mathematical circles of correspondence* of which P. de Carcany and M. Mersenne were the intermediaries,

* These circles were the predecessors of the scientific societies of to-day, and were the normal means of spreading results at this period.

many definite integrations were exchanged. Notably we may mention Pascal (1623–62), who used the method in 1659 to find the area, volume of revolution and centre of gravity of the cycloid, and Wallis (1616–1703), a Cambridge mathematician who migrated to Oxford as Savilian Professor of Geometry. He expressed a complete set of rules for the summation of the series $\Sigma x^n dx$ for all cases except $n = -1$. His celebrated formulæ mentioned in Ch. V were obtained by regarding the area of a circle $\{y = (x - x^2)^{\frac{1}{2}}\}$ as intermediate between that of the curves $y = (x - x^2)^0$, $y = (x - x^2)^1$. A few years later he also gave some formulæ for the rectification of plane curves. Many other mathematicians of this period effected summations of this type, amongst whom we may mention Fermat, Edward Wright (who had a formula equivalent to $\int \sec \theta d\theta$), Roberval, Huyghens, and Torricelli.

We thus see that up to 1660 mathematicians had a good idea of the far-reaching results that might be effected by a general method of summation, and could perform many individual summations, but had no general rule for performing them. The discovery of the Theory of Fluxions by Newton and the Differential Calculus by Leibnitz at about the same date immediately linked up these results in the manner described in Ch. VI on the Fundamental Theorem. Almost immediately, therefore, a complete Integral Calculus sprang into being; the dry bones had long been there, this was the breath that gave them life. Space prohibits us from going into the long and bitter controversy which raged for over 100 years around the two principal figures in these discoveries, and which in some respects had a most unfortunate effect on English mathematics. Suffice it, however, to say that Newton had used his methods in the tract "de Quadratura Curvarum" many years before they were published—and that he knew the outline of the work at any rate as early as 1666. We do know definitely that Leibnitz saw some MSS of Newton's, during his visit to London in 1673, and quite possibly again in 1675 on a visit to Tschirnhausen, but we have no very clear evidence as to what ideas he derived from them. It is only fair to say that the controversy was not sought by the principal figures but by some of their followers, and it is unpleasant to have to add that many of Newton's supporters were more distinguished by their zeal than by their discretion or skill. Only one advantage arose out of this most unfortunate episode; I refer to the "challenge problems" sent out by both sides —at first merely as tests in the efficacy of their own methods, and later as annoyances to the other side. As these problems were often solved by both parties, they contributed but little to the controversy. They have however contributed many results, notably in Dynamics and Kinematics, which remain in the text-books to this day.

There is the other sad side to this subject. The main result was a standing aside of the English mathematicians from the results achieved on the Continent and a persistence in using the old notation and methods, which held English mathematics back at least 100 years, and from which we have hardly even now recovered. It was actually not till 1819 that Peacock, Herschel, and Babbage (the former of whom was moderator that year) were able to introduce the continental notation—or as they themselves cleverly put it* "to advocate the principles of pure

---

* In the Newtonian notation $\dot{x}$ is used where the continental writes $\dfrac{dx}{dt}$.

'd'-ism instead of the 'dot'-age of the University." So that finally the nett result was a grafting of the Leibnitzian notation on to Newtonian ideas.

However, four of Newton's contemporaries and immediate successors made some contributions to the theory. Brook Taylor (1685-1731) introduced the idea of changing the variable (Ch. III). Partial Fractions were studied by Cotes (1682-1716), whose early death robbed England of a second Newton, and his work was completed by de Moivre; whilst a standard treatise on the whole method was written by Maclaurin in 1742.

But in the main the development of the subject on the Continent was far in advance of that in this country, owing to the superiority of the notation employed. Leibnitz himself had invented the symbols $\int$ (at first he wrote "omnia y" instead of $\int y$) and $\frac{d}{dx}$, and the methods were enthusiastically supported by the brothers Bernoulli (James and John) who were successively professors at Bâle. James Bernoulli wrote in 1691 the first treatise on Integral Calculus, Count J. F. Riccati spread the new doctrines in Italy, and d'Alembert (1717-83) in his "Traité de dynamique" showed how the principles governing the motion of a body could be expressed by the new notation. The subject of Differential Equations naturally grew up at the same time, but its development has taken much longer and in fact is not yet finally completed. The theory of the simple equations given in this book is however mainly due to d'Alembert and Leonhard Euler (1707-83). The latter made so many valuable contributions to all branches of mathematical research that it is hardly possible to find any branch of the subject which has not a theorem bearing his name associated with it. He published a vast quantity of works, some of which are still read, including (in 1770) his "Institutiones Calculi Integralis," a very full account of the theory, including the method of successive reductions and the Beta and Gamma Functions. The properties of Resonance, which are mentioned in the last chapter, were studied by Lagrange, who published in 1772–85 a series of tracts on Differential Equations which practically forms the basis of the whole subject as read to-day.

The student who is desirous of further historical information is referred to the standard works, particularly those by the late W. W. R. Ball and Prof. F. Cajori, to which the author himself is mainly indebted. It is the author's opinion that far too little is learnt about those mighty men who laid the foundations of our knowledge, and that, if we studied their history more, we might catch perchance a little of the great enthusiasm and inspiration with which they made their immortal discoveries.

CHAPTER I
DEFINITIONS AND STANDARD FORMS
1. Definition of Integration.

Integration is defined as the reversal of the process of differentiating. In the simpler cases one variable only is employed. Hence the process of "integrating" a given function of x is equivalent to that of finding a second function of x, which is such that the first function is the "differential" or "derivative" of the second. Or to put the matter into mathematical language—to integrate $F(x)$ is to find another function $\phi(x)$, such that $\frac{d}{dx}\phi(x) = F(x)$.

There is a standard notation for this, which is now almost universally adopted, viz. $\int F(x)\,dx$—which should be read "the integral of $F(x)$ with respect to x." The symbol $\int$ is in reality a conventionalized capital S and signifies the fact that integration is connected with the process of summation, as we shall see in Chapter VI.

2. Character of Integration.

A consideration which strikes us at the outset, is that in the final resort integration is a species of guesswork, i.e. we have to be able to see by intuition the kind of result which will, when differentiated, produce the function given. But it is not a process of illogical or unreasoning guesswork. The reader will find that there are definite things to try and definite rules to follow and to the total collection of these rules is given the name of "Integral Calculus."

We also remember that the differential of a constant is

zero. It follows, then, that when we try to reverse the process of differentiating we do not know whether to add a constant to the result or not, or, in any case, what constant to add.

For instance $\dfrac{d}{dx}(x^4)$, $\dfrac{d}{dx}(x^4 + 4)$ and $\dfrac{d}{dx}(x^4 + 100)$ are all equal to $4x^3$. Which of these functions then is the "integral" of $4x^3$? We get over the difficulty by saying that $\int 4x^3\,dx$ is $x^4 + C$, where C is a constant that remains to be determined or, as we usually put it, "an arbitrary constant." In any practical question, sufficient data are usually given to determine the value of this constant.

3. Indefinite Integrals. Necessity for Inclusion of Constant.

When an integral is given in this form it is known as an "indefinite" integral. The arbitrary constant should always be included unless it is quite clearly non-existent; its omission can often lead to serious error, as the following example will show.

We remember from Differential Calculus that

$$\frac{d}{dx}(-\cos^{-1}x) \text{ and } \frac{d}{dx}(\sin^{-1}x) \text{ are both equal to } \frac{1}{\sqrt{1 - x^2}}.$$

It follows then (if we omit the arbitrary constants) that $\sin^{-1}x$ and $-\cos^{-1}x$ are equal, or that $\sin^{-1}x + \cos^{-1}x = 0$, a result which is plainly ridiculous.

What we should have written is

$$\text{(i)} \quad \int\frac{1}{\sqrt{1 - x^2}} = \sin^{-1}x + C,$$

or
$$\text{(ii)} \quad \int\frac{1}{\sqrt{1 - x^2}} = -\cos^{-1}x + D;$$

whence $\sin^{-1}x + C = -\cos^{-1}x + D,$

or $\sin^{-1}x + \cos^{-1}x = D - C.$

This result is not impossible since we know that

$$\sin^{-1}x + \cos^{-1}x = \frac{\pi}{2},$$

and we have merely to choose the constants D and C so that they differ by $\frac{\pi}{2}$.

4. Since, as we have seen, integration and differentiation are reverse processes, it follows that the integral of the sum of a number of separate functions is the sum of the integrals of the separate functions,

viz. $\int (u+v+w+\ldots)dx = \int u\,dx + \int v\,dx + \int w\,dx + \ldots.$

It is also obvious that integration is commutative with respect to a constant multiplier,

i.e. if A is constant $\int A u\,dx = A \int u\,dx.$

A table of Standard Forms can be constructed by reversing well-known results of Differential Calculus. Such a table is given here but the reader will find it of far greater value to construct his own table and keep it for reference, adding to it from time to time any new results he may meet. The results are given here both in the Differential and Integral Forms but in all cases the arbitrary constant has, for considerations of space, been left out. It is, of course, much more essential that such a table should exist in the reader's memory than in the text-book, as a thorough knowledge of it is essential to his further progress.

5. Table of Standard Forms.

Differential Calculus	*Integral Calculus*

$$\frac{d}{dx} x^n = nx^{n-1}.$$

$$\int x^n dx = \frac{x^{n+1}}{n+1} \text{ unless } n = -1.$$

$$\frac{d}{dx}(ax+b)^n = na(ax+b)^{n-1}.$$

$$\int (ax+b)^n dx = \frac{(ax+b)^{n+1}}{(n+1)a}.$$

$$\frac{d}{dx} \log_e x = \frac{1}{x}.$$

$$\int \frac{1}{x} dx = \log_e x.$$

$$\frac{d}{dx} e^{ax} = ae^{ax}.$$

$$\int e^{ax} dx = \frac{1}{a} e^{ax}.$$

$$\frac{d}{dx} \sin x = \cos x.$$

$$\int \cos x\, dx = \sin x.$$

$$\frac{d}{dx} \cos x = -\sin x.$$

$$\int \sin x\, dx = -\cos x.$$

$$\frac{d}{dx} \sin mx = m \cos mx.$$

$$\int \cos mx\, dx = \frac{1}{m} \sin mx.$$

$$\frac{d}{dx} \tan x = \sec^2 x.$$

$$\int \sec^2 x\, dx = \tan x.$$

$$\frac{d}{dx} \cot x = -\operatorname{cosec}^2 x.$$

$$\int \operatorname{cosec}^2 x\, dx = -\cot x.$$

$$\left.\begin{array}{l} \dfrac{d}{dx} \sin^{-1}\dfrac{x}{a} = \dfrac{1}{\sqrt{a^2-x^2}} \\[2ex] \dfrac{d}{dx} \cos^{-1}\dfrac{x}{a} = -\dfrac{1}{\sqrt{a^2-x^2}} \end{array}\right\}.$$

$$\int \frac{dx}{\sqrt{a^2-x^2}} = \begin{cases} \sin^{-1}\dfrac{x}{a}, \\[2ex] -\cos^{-1}\dfrac{x}{a}. \end{cases}$$

$$\frac{d}{dx} \tan^{-1}\frac{x}{a} = \frac{a}{x^2+a^2}.$$

$$\int \frac{dx}{a^2+x^2} = \frac{1}{a} \tan^{-1}\frac{x}{a}$$
$$\left(\text{or } -\frac{1}{a}\cot^{-1}\frac{x}{a}\right).$$

$$\frac{d}{dx} a^x = a^x \log_e a.$$

$$\int a^x dx = \frac{a^x}{\log_e a}.$$

N.B. (1) Notice that all integrals commencing with the prefix "co-" have a negative sign prefixed.

(2) In cases like $\int \dfrac{dx}{a^2+x^2}$ and $\int \dfrac{dx}{\sqrt{a^2-x^2}}$ it is useful to notice the

dimensions: for instance $\int \dfrac{dx}{a^2 + x^2}$ is of dimensions -1, but $\tan^{-1}\dfrac{x}{a}$ is of zero dimensions, hence we clearly must put in a factor $\dfrac{1}{a}$ of dimensions -1; on the other hand $\int \dfrac{dx}{\sqrt{a^2 - x^2}}$ and $\sin^{-1}\dfrac{x}{a}$ are both of zero dimensions and no factor is required.

(3) It is interesting to notice that the case $\int \dfrac{1}{x}\,dx$ is not really an exception to the rule for $\int x^n\,dx$.

For
$$\int x^n\,dx = \frac{x^{n+1}}{n+1} + A,$$

or as we may equally put it

$$= \frac{x^{n+1} - 1}{n+1} + B,$$

where B is another arbitrary constant.

Now consider the Lt as $n \to -1$.

Then
$$\int x^{-1}\,dx = \operatorname*{Lt}_{n \to -1} \frac{x^{n+1} - 1}{n+1} + B$$
$$= \operatorname*{Lt}_{m \to 0} \frac{x^m - 1}{m} + B,$$

putting $m = n + 1$.

But by a Theorem on Limits (Chrystal's *Algebra*, ch. xxv, § 13)

$$\operatorname*{Lt}_{m \to 0} \frac{x^m - 1}{m} = \log x.$$

$$\therefore \quad \int x^{-1}\,dx = \log x + B.$$

EXAMPLES I A

Integrate the following:

(1) $\sqrt{x}, \quad \dfrac{1}{\sqrt{x}}, \quad \dfrac{1 + x + x^2}{3}$.
(2) $\sqrt[3]{x^4}, \quad \dfrac{1}{x^6}, \quad \dfrac{1}{\sqrt[5]{x^7}}$.

(3) $\sqrt[3]{3x - 4}, \quad \sqrt{3x - 4}, \quad \dfrac{1}{\sqrt{3x - 4}}$.

(4) $\dfrac{1}{a + x}, \quad \dfrac{1}{a + x} + \dfrac{1}{a - x}, \quad \dfrac{1}{(a - x)^n}$.

(5) $\dfrac{1}{x+1}$, $\dfrac{x+3}{x+1}$, $\dfrac{1}{ax+b}$, $\dfrac{ax+c}{ax+b}$, $\dfrac{Ax+B}{ax+b}$.

(6) $\dfrac{1+x}{\sqrt{x}}$, $\dfrac{x^2+1}{x}$, $\dfrac{1}{\sqrt{x+1}+\sqrt{x}}$, $x\sqrt{1+x}$.

(7) Verify by differential that $\displaystyle\int\dfrac{dx}{\sqrt{x^2+a^2}}=\log(x+\sqrt{x^2+a^2})+c$.

(8) Find by trial, noticing result of Ex. 7, $\displaystyle\int\dfrac{dx}{\sqrt{x^2-a^2}}$.

6. Two Special Functional Forms.

Remembering that $\dfrac{d}{dx}\log\{\phi(x)\}=\dfrac{\phi'(x)}{\phi(x)}$, it is clear that

$$\int\frac{\phi'(x)\,dx}{\phi(x)}=\log\phi(x)+C.$$

Or putting the matter into words: "the integral of a fraction whose numerator is the differential of its denominator is the logarithm of the denominator." We shall have occasion to use this form in Ch. IV, and the rule is of frequent general application.

We see also that

$$\int[\phi(x)]^n\cdot\phi'(x)\,dx=\frac{1}{n+1}[\phi(x)]^{n+1}+C.$$

A few examples of the applications of these rules are given:

(1) $\displaystyle\int\cot x\,dx=\int\dfrac{\cos x}{\sin x}\,dx.$

Here $\phi(x)=\sin x$.

Hence $\displaystyle\int\cot x\,dx=\log\sin x+C.$

Again

$$\int\tan x=-\log\cos x+C \text{ or } \log\sec x+C.$$

(2) Similarly $\displaystyle\int\dfrac{2ax+b}{ax^2+bx+c}\,dx=\log(ax^2+bx+c)+C.$

(3) $\int \dfrac{\cos x}{\sin^2 x} dx = \int (\sin x)^{-2} . \cos x \, dx.$

Here $\phi(x) = \sin x.$

$$\therefore \int \frac{\cos x}{\sin^2 x} dx = -\frac{1}{\sin x} = -\operatorname{cosec} x.$$

Or again $\int \sin^4 x \cos x \, dx = \dfrac{\sin^5 x}{5} + C.$

Lastly to find $\int \tan^3 x \, dx.$

$$\int \tan^3 x \, dx = \int \tan x . \tan^2 x \, dx$$

$$= \int \tan x \, (\sec^2 x - 1) \, dx$$

$$= \int \tan x \sec^2 x \, dx - \int \tan x \, dx$$

$$= \frac{1}{2} \tan^2 x + \log \cos x \,;$$

this is seen to be an example of the use of both rules.

The student should observe that these devices are merely methods of representing the reversal of the rule for differentiating a "function of a function"; and if the student has become skilled in differentiating expressions of this sort he will have little difficulty in integrating, or in recognising how a given expression might be derived.

In some of the examples following, it may be necessary to employ some transformation before commencing to integrate. This is often the case with trigonometrical functions such as $\cos^2 x$, etc., which should always be expressed in terms of double the angle.

EXAMPLES I B

Integrate the following:

(1) $2 \sin x \cos x$, $\cos^2 x$, $\cos mx \cos nx$, $\sin mx \sin nx$.

(2) $\tan x$, $\dfrac{\sin x}{\cos^2 x}$, $\cos x \left\{ \dfrac{1}{\sin x} + \dfrac{1}{\sin^3 x} \right\}$, $\dfrac{\sec^2 x}{\tan x}$.

(3) $\dfrac{1}{x \log x}$, $\dfrac{e^x}{e^x + 2}$, $\dfrac{e^x + e^{-x}}{e^x - e^{-x}}$, $\dfrac{\cot x}{\log \sin x}$.

(4) $\dfrac{2x - 3}{x^2 - 3x + 2}$, $\dfrac{2x + 1}{2x - 1}$, $(x^2 + 2x + 2)(x + 1)$, $\dfrac{x^2 + 2x + 2}{x + 1}$, $\dfrac{x + 1}{x^2 + 2x + 2}$.

(5) $\dfrac{1}{x^2 + 9}$, $\dfrac{x^2}{x^2 + 4}$, $\dfrac{2x}{1 + x^4}$, $\dfrac{1}{7 + 3x^2}$.

(6) $\cos^3 x \sin x$, $\sin^2 x \cos^3 x$, $\cos^3 x$.

CHAPTER II

INTEGRATION BY SUBSTITUTION,
OR
CHANGING THE SUBJECT OF INTEGRATION

7. Definite Integrals.

We have already spoken of indefinite integrals in the last chapter. Before proceeding further we ought to define what is meant by a "definite integral."

The expression $\int_a^b f(x)\, dx$ is called " the definite integral of $f(x)$ between the limits of integration b and a." Its value is formed by first calculating the indefinite integral of $f(x)$ and then subtracting the two values of this expression when $x = b$ and when $x = a$, from one another, the value for the upper limit being always taken first.

For example, to find $\int_a^b x^2 dx$,

$$\int x^2\, dx = \frac{x^3}{3} + C,$$

$$\therefore \text{ definite integral } \int_a^b x^2\, dx = \left[\frac{x^3}{3} + C\right]_a^b$$

$$= \frac{b^3}{3} + C - \frac{a^3}{3} - C$$

$$= \frac{b^3 - a^3}{3}.$$

We notice that in a definite integral the arbitrary constant always disappears, and consequently it is usually omitted.

We shall see later in Ch. VI that a physical meaning can often be assigned to a Definite Integral, and it is this type of integral which makes its appearance in questions connected with Geometry, Mechanics, etc.

8. Changing the subject of Integration.

As we have noticed, Integration is, in the last resort, a species of guesswork. It very often happens, however, that the function to be integrated is not expressed in a form which allows us to see readily of what function it is the differential. A large part of the theoretical side of the Integral Calculus consists of various devices by which we can reduce an apparently difficult integral to a form in which it can be more readily recognized.

The first and most important of these processes is that of Substitution or "Changing the subject of Integration" and it is very important that the student should master this thoroughly before going any further.

Suppose that we have an integral, $\int V\,dx$, which appears to be insoluble as it stands, but that we can see that the form of the expression V can be simplified if, instead of some function of x, we write some other variable z (such as, for instance, $z = \cos x$, or $z = e^x$).

Now let $u = \int V\,dx$.

Suppose that when we put z for x the expression V changes into a new expression V_z.

Now by the definition of an integral

$$\frac{du}{dx} = V.$$

But since z and x are connected

$$\frac{du}{dx} = \frac{du}{dz} \cdot \frac{dz}{dx}.$$

$$\therefore \frac{du}{dz} = V \frac{dx}{dz}.$$

The right-hand side of this equation can now be expressed entirely in terms of z and we get

$$\frac{du}{dz} = V_z \frac{dx}{dz}.$$

$$\therefore u = \int V_z \frac{dx}{dz} \, . \, dz.$$

We thus see that we can, without altering the result, change the variable from x to z provided we insert instead of dx the expression $\dfrac{\mathbf{dx}}{\mathbf{dz}}$ dz. When the result is written entirely in terms of z, it may be easier to integrate than the original expression in terms of x.

9. If the original integral is indefinite, the result must be converted back so that it appears in terms of x when the integration is completed. If however the integral is a definite integral, we can avoid this process, which is often laborious, by changing the limits. We take as the new limits the value of z corresponding to the limiting values of x, and the definite integral can then be found without reverting to the original variable. The choice of the right substitution often demands considerable ingenuity, and the whole method is best illustrated by worked examples, which we give in the following section.

Many new standard forms can be found in this way, and the student is advised to add the results printed in heavier type to his table of Standard Forms.

10. Worked Examples.

(i) We first choose an example of which we know the result,

$$\text{viz.}\quad \int \frac{1}{\sqrt{1-x^2}}\,dx.$$

It is seen that if we put $x = \sin\theta$ the radical disappears, giving $\cos\theta$ in the denominator.

The work is usually arranged as follows:

Let $x = \sin\theta$; then $\quad\dfrac{dx}{d\theta} = \cos\theta,$

i.e. for dx we put $\cos\theta \cdot d\theta$.

$$I = \int \frac{1}{\sqrt{1-\sin^2\theta}} \cdot \cos\theta \cdot d\theta$$

$$= \int \frac{1}{\cos\theta} \cdot \cos\theta \cdot d\theta$$

$$= \int 1\,d\theta = \theta + C.$$

But $\theta = \sin^{-1}x$.

Hence $\qquad \int \dfrac{dx}{\sqrt{1-x^2}} = \sin^{-1}x + C.$

If we had chosen the substitution $x = \cos\theta$ we should have arrived at the other form of the integral which has already been noted in Ch. I, viz. $-\cos^{-1}x + D$.

(ii) $\int \sqrt{a^2 - x^2}\,dx.$

Again put $\quad x = a\sin\theta$, then $\dfrac{dx}{d\theta} = a\cos\theta$

or $\qquad\qquad\qquad\qquad dx = a\cos\theta \cdot d\theta.$

$$I = \int \sqrt{a^2 - a^2 \sin^2 \theta} \, . \, a \cos \theta \, . \, d\theta$$

$$= \int a^2 \cos^2 \theta \, . \, d\theta$$

$$= a^2 \int \frac{1 + \cos 2\theta}{2} \, d\theta$$

$$= \frac{a^2}{2} \int 1 d\theta + \frac{a^2}{2} \int \cos 2\theta d\theta$$

$$= \frac{a^2 \theta}{2} + \frac{a^2 \sin 2\theta}{4}$$

$$= \frac{a^2}{2} [\theta + \sin \theta \cos \theta]$$

$$= \frac{a^2}{2} \left[\sin^{-1} \frac{x}{a} + \frac{x}{a} \sqrt{1 - \frac{x^2}{a^2}} \right]$$

$$= \frac{a^2}{2} \sin^{-1} \frac{x}{a} + \frac{x \sqrt{a^2 - x^2}}{2} .$$

Hence $\quad \int \sqrt{a^2 - x^2} dx = \dfrac{a^2}{2} \sin^{-1} \dfrac{x}{a} + \dfrac{x \sqrt{a^2 - x^2}}{2} .$

(iii) Integrals containing the expression $(x^2 + a^2)$ can often be simplified by putting $x = a \tan \theta$.

Let us try $\quad \displaystyle\int_0^a \frac{x}{\sqrt{x^2 + a^2}} dx.$

Let $x = a \tan \theta$, i.e. $dx = a \sec^2 \theta d\theta$.

Also when $x = a$, $\theta = \dfrac{\pi}{4}$; when $x = 0$, $\theta = 0$.

Hence the new limits are $\dfrac{\pi}{4}$ and 0.

$$\therefore I = \int_0 \frac{a \tan \theta}{\sqrt{a^2 + a^2 \tan^2 \theta}} \cdot a \sec^2 \theta \, d\theta$$

$$= \int_0^{\overline{4}} a \tan \theta \sec \theta \, d\theta$$

(remembering $\sec^2 \theta = 1 + \tan^2 \theta$)

$$= a \int_0^{\frac{\pi}{4}} \frac{\sin \theta}{\cos^2 \theta} \, d\theta$$

$$= a \left[\frac{1}{\cos \theta} \right]_0^{\frac{\pi}{4}} = a \, (\sqrt{2} - 1).$$

(iv) Algebraic substitutions are sometimes useful.

Consider $\int \dfrac{dx}{2\sqrt{x} \cdot (1 + x)}$.

Put $\sqrt{x} = z$; i.e. $\dfrac{1}{2\sqrt{x}} \, dx = dz$,

$$I = \int \frac{dz}{1 + z^2}$$

$$= \tan^{-1} z + C$$

$$= \tan^{-1} \sqrt{x} + C.$$

Note. A few further suggestions are appended for integrals of a more difficult character.

(1) Any expression containing a single linear surd, e.g. $\sqrt{ax + b}$, can be reduced to an algebraic form by the substitution $\sqrt{ax + b} = t$.

The integration may then be effected by the methods given and those of the following chapters.

(2) An expression containing a double surd of the form $\sqrt{(a - x)(x - b)}$ can be treated by the substitution

$x = a \cos^2 \theta + b \sin^2 \theta$, which renders it trigonometric and integrable as in Ch. V.

(3) Any expression containing a surd of the form $\sqrt{(x-a)(x-b)(x-c)}$ is not integrable in terms of known functions. Such an expression in fact gives rise to Elliptic Functions.

EXAMPLES II

Integrate by the method of substitution:

(1) $\sqrt{1-x}$.   (2) $\sqrt[3]{1+2x}$.   (3) $\dfrac{2x}{1+x^4}$.   (4) $\dfrac{3x^2}{1+x^6}$.

(5) $\dfrac{1}{x^2+2x+3}$ $\{z = x+1\}$.   (6) $\dfrac{1}{\sqrt{8-2x-x^2}}$.

(7) $\dfrac{1}{(x-1)\sqrt{x^2-2x}}$.   (8) $\dfrac{1}{\sqrt{x-x^2}}$ $\{z = x - \frac{1}{2}\}$.

(9) $\dfrac{1}{e^x+e^{-x}}$.   (10) $e^x \cos e^x$.   (11) Evaluate $\displaystyle\int_0^a \dfrac{x}{\sqrt{a^2-x^2}}\, dx$.

(12) Evaluate $\displaystyle\int_0^a x^2 \sqrt{a^3+x^3}\, dx$.

(13) By writing $z = \tan \dfrac{x}{2}$, prove that

$$\int \mathbf{cosec\ x\, dx} = \mathbf{\log \tan \dfrac{x}{2} + C}$$

and

$$\int \mathbf{sec\ x\, dx} = \mathbf{\log \tan \left(\dfrac{\pi}{4} + \dfrac{x}{2}\right) + C,}$$

or

$$\mathbf{\log (sec\ x + \tan x) + C.}$$

(14) Use the last example to prove that

$$\int \dfrac{\mathbf{dx}}{\sqrt{\mathbf{a^2+x^2}}} = \mathbf{\log \dfrac{x+\sqrt{x^2+a^2}}{a} + C.}$$

(15) Using result of Ex. 14, prove that

$$\int \sqrt{\mathbf{a^2+x^2}}\, \mathbf{dx} = \dfrac{\mathbf{x\sqrt{a^2+x^2}}}{\mathbf{2}} + \dfrac{\mathbf{a^2}}{\mathbf{2}} \mathbf{\log \left(\dfrac{x+\sqrt{x^2+a^2}}{a}\right) + C.}$$

Find the following:

(16) $\int \dfrac{dx}{1+\cos x}$.   (17) $\int \dfrac{dx}{x\sqrt{2ax-x^2}}$.   (18) $\int \dfrac{dx}{x\sqrt{x^2-a^2}}$.

(19) $\int \dfrac{dx}{\sqrt{2ax-x^2}}$.   *(20) $\int (\tan x + \tan^3 x)\,dx$.

*(21) $\displaystyle\int_0^{\frac{1}{\sqrt{2}}} x^2 \sqrt{1-x^2}\,dx$.

*(22) Show that $\int \dfrac{dx}{\sqrt{(x-a)(x-b)}}$ can be evaluated by the substitution $y^2 = \dfrac{x-a}{x-b}$.

*(23) Show that the integrals $\int \dfrac{1-x^2}{1+kx^2+x^4}\,dx$ and $\int \dfrac{1+x^2}{1+kx^2+x^4}\,dx$ can be found by the substitutions $v = x - \dfrac{1}{x}$, $u = x + \dfrac{1}{x}$.

*(24) By using the substitution $x+1 = \dfrac{1}{y}$, prove that

$$\int_1^3 \dfrac{dx}{(x+1)\sqrt{4x-3-x^2}} = \dfrac{\pi}{2\sqrt{2}}.$$

CHAPTER III

INTEGRATION BY PARTS

11. We have noticed already that the art of integration lies in transforming a difficult integral into one which can be recognized more readily as a known form. When the function to be integrated is in the form of a product, we can sometimes effect this by the device known as "Integration by Parts." This is the result in Integral Calculus which corresponds to the rule for differentiating a product and is easily found as follows.

We know that if u and w are functions of x, then

$$\frac{d}{dx}\{uw\} = w \cdot \frac{d}{dx}(u) + u \cdot \frac{d}{dx}(w).$$

Now if we integrate both sides of this equation we get

$$uw = \int w \cdot \frac{du}{dx} \cdot dx + \int u \cdot \frac{dw}{dx} \cdot dx \ldots\ldots\ldots(A).$$

We can for the sake of convenience consider the arbitrary constants as included in the integrals.

Now u and w are perfectly general and unrestricted functions. Let us suppose that w is the integral of such other function v, i.e. that

$$w = \int v\,dx$$

or

$$\frac{dw}{dx} = v.$$

Making this substitution in equation (A) we now have

$$u \cdot \int v\,dx = \int \left[\left\{ \int v\,dx \right\} \cdot \frac{du}{dx} \right] dx + \int uv\,dx,$$

or rearranging

$$\int uv\,dx = u \cdot \int v\,dx - \int \left(\int v\,dx \right) \cdot \frac{du}{dx} \cdot dx.$$

The result is perhaps more clearly stated in words as follows:

Integral of product = first function × Integral of second − Integral of {Differential of first × Integral of second}.

12. We notice, first of all, that the method does not "evaluate" the integral of a product; there is, in fact, no general formula to do this; what it does enable us to do is to connect the integral of a product with a second integral which may be easier than the original. The rule is, in fact, the simplest instance of a Reduction Formula, of which we shall have further examples in Ch. V. Care must, of course, be taken to choose the order of the functions in such a way as to reduce the second integral as far as possible. Otherwise we shall gain nothing by the transformation. In certain examples it may be necessary to apply the process more than once before we finally arrive at a result which can be integrated at sight. The necessity for a right choice of the order of the functions is illustrated by the following parallel attempts to evaluate the same integral:

$$\int x \cos x\,dx.$$

Let $u = x$, $\quad v = \cos x$,

$\dfrac{du}{dx} = 1$, $\quad \int v\,dx = \sin x$.

$I = x \cdot \sin x - \int 1 \cdot \sin x\,dx$

$\quad = \mathbf{x \sin x + \cos x}.$

$$\int x \cos x\,dx.$$

Let $u = \cos x$, $\quad v = x$,

$\dfrac{du}{dx} = -\sin x$, $\quad \int v\,dx = \dfrac{x^2}{2}$.

$I = \dfrac{x^2}{2} \cos x + \int \dfrac{x^2}{2} \sin x\,dx$,

a result which is quite true but does not help us.

13. Worked Examples.

Careful note should be taken of the following worked examples:

(i) $\int x e^x\, dx.$
$\qquad\qquad\qquad$ Let $u = x,\qquad v = e^x,$

$$\frac{du}{dx} = 1, \qquad \int v\, dx = e^x$$

$$\therefore\ I = x\,.\,e^x - \int 1\,.\,e^x\, dx$$

$$= x e^x - e^x + C$$

$$= (x - 1)\, e^x + C.$$

(ii) In some cases unity may be used as a function,

e.g. $\int \log x\, dx = \int 1\,.\,\log x\, dx.$ $\qquad$ Let $u = \log x,\qquad v = 1,$

$$\frac{du}{dx} = \frac{1}{x}, \qquad \int v\, dx = x.$$

$$\therefore\ I = x \log x - \int 1\,.\,dx$$

$$= x \log x - x + C$$

$$= x\,(\log x - 1) + C$$

or $\qquad\qquad\qquad x \log \dfrac{x}{e} + C.$

(iii) The process may have to be repeated two or more times before the evaluation can be effected.

Consider $\int x^2 \sin x\, dx.$ $\qquad$ Here $u = x^2,\qquad v = \sin x,$

$$\frac{du}{dx} = 2x, \quad \int v\, dx = -\cos x.$$

$$\therefore\ I = - x^2 \cos x + \int 2x \cos x\, dx.$$

To evaluate $\int x \cos x \, dx$ we repeat the process, getting the result $x \sin x + \cos x + C$ as in § 12.

Hence finally

$$I = -x^2 \cos x + 2x \sin x + 2 \cos x + C$$
$$= -\cos x \, (x^2 - 2) + 2x \sin x + C.$$

14. An Important Artifice.

We can sometimes by a little ingenuity bring the second integral back to the same form as the original integral, and thus find the value of the integral directly.

This is a very useful and important artifice and the example given should be carefully studied.

To find $\int \sqrt{a^2 + x^2} \, dx.$ Let $u = \sqrt{a^2 + x^2}$, $v = 1$,

$$\frac{du}{dx} = \frac{x}{\sqrt{a^2 + x^2}}, \quad \int v \, dx = x.$$

Then
$$I = x \sqrt{a^2 + x^2} - \int \frac{x^2}{\sqrt{a^2 + x^2}} \, dx$$

$$= x \sqrt{a^2 + x^2} - \int \left\{ \frac{x^2 + a^2}{\sqrt{a^2 + x^2}} - \frac{a^2}{\sqrt{a^2 + x^2}} \right\} dx$$

$$= x \sqrt{a^2 + x^2} - \int \sqrt{a^2 + x^2} \, dx + a^2 \int \frac{dx}{\sqrt{a^2 + x^2}}$$

$$= x \sqrt{a^2 + x^2} - I + a^2 \log \frac{x + \sqrt{x^2 + a^2}}{a} + C,$$

cf. Ex. II, 14.

$$\therefore \; 2I = x \sqrt{a^2 + x^2} + a^2 \log \frac{x + \sqrt{x^2 + a^2}}{a},$$

i.e. $\displaystyle\int \sqrt{a^2 + x^2} \, dx = \frac{x \sqrt{a^2 + x^2}}{2} + \frac{a^2}{2} \log \frac{x + \sqrt{a^2 + x^2}}{a} + C.$

This method is also applicable to

$$\int \sqrt{x^2 - a^2}\, dx, \quad \int \sqrt{a^2 - x^2}\, dx,$$

and the student should perform the process for these forms and make a careful note of the result.

15. The forms $e^{ax} \cos bx$, $e^{ax} \sin bx$.

These forms can at once be integrated by this method, two successive applications being necessary.

Let
$$I = \int e^{ax} \sin bx\, dx.$$

Integrate by parts, then

$$I = -e^{ax}\frac{\cos bx}{b} + \int \frac{a}{b} e^{ax} \cos bx\, dx.$$

But again

$$\int e^{ax} \cos bx\, dx = e^{ax}\frac{\sin bx}{b} - \int \frac{a}{b} e^{ax} \sin bx\, dx.$$

Therefore substituting

$$I = -e^{ax}\frac{\cos bx}{b} + \frac{a}{b^2} e^{ax} \sin bx - \frac{a^2}{b^2} I,$$

i.e.
$$\frac{a^2 + b^2}{b^2} I = \frac{ae^{ax} \sin bx - be^{ax} \cos bx}{b^2},$$

or
$$I = \frac{e^{ax}}{a^2 + b^2}\{a \sin bx - b \cos bx\} + C$$

$$= \frac{e^{ax}}{\sqrt{a^2 + b^2}}\left\{\frac{a}{\sqrt{a^2 + b^2}} \sin bx - \frac{b}{\sqrt{a^2 + b^2}} \cos bx\right\} + C.$$

Now let $\tan \alpha = \dfrac{b}{a}$, then

$$\frac{a}{\sqrt{a^2 + b^2}} = \cos \alpha \ \text{ and } \ \frac{b}{\sqrt{a^2 + b^2}} = \sin \alpha.$$

Hence

$$I = \frac{e^{ax}}{\sqrt{a^2 + b^2}} \{\sin bx \cos \alpha - \cos bx \sin \alpha\} + C$$

$$= \frac{e^{ax}}{\sqrt{a^2 + b^2}} \sin (bx - \alpha) + C$$

$$= \frac{1}{\sqrt{a^2 + b^2}} e^{ax} \sin \left(bx - \tan^{-1} \frac{b}{a}\right) + C^*.$$

Similarly we can prove that

$$\int e^{ax} \cos bx \, dx = \frac{1}{\sqrt{a^2 + b^2}} e^{ax} \cos \left(bx - \tan^{-1} \frac{b}{a}\right) + C'^*.$$

Ex. Integrate $e^{3x} \sin 4x \cos 6x$.

$$I = \frac{1}{2} \int e^{3x} (\sin 10x - \sin 2x) \, dx,$$

$$2I = \int e^{3x} \sin 10x \, dx - \int e^{3x} \sin 2x \, dx$$

$$= \frac{e^{3x}}{\sqrt{109}} \sin \left(10x - \tan^{-1} \frac{10}{3}\right) - \frac{e^{3x}}{\sqrt{13}} \sin \left(2x - \tan^{-1} \frac{2}{3}\right) + C.$$

EXAMPLES III

Integrate the following :

(1) $xe^{-x}, \ xe^{2x}, \ x^2 e^x, \ x^3 e^x$.

(2) $x \sin x, \ x^2 \cos x, \ x^3 \sin x, \ x \cos^2 x, \ x \sin x \cos x$.

(3) $x \log x, \ x^2 \log x, \ x^n \log x$.

(4) $e^x \cos x, \ e^x \cos 2x, \ e^x \sin 2x, \ e^x (\sin x + \cos x)$.

(5) $\sin^{-1} x, \ \tan^{-1} x, \ x \sin^{-1} x, \ x \tan^{-1} x$.

(6) $\sqrt{x^2 - a^2}, \ \sqrt{a^2 - x^2}$.

(7) $e^{3x} \cos 4x, \ 2e^{5x} \sin 7x \cos 3x, \ 4e^x \sin x \sin 2x \sin 3x$.

(8) $x \sec^2 x, \ \dfrac{\sin^{-1} x}{(1 - x^2)^{\frac{3}{2}}}$.

* This can be directly deduced from the rule for differentiating these functions given in the standard books on Differential Calculus.

CHAPTER IV

RATIONAL ALGEBRAIC FRACTIONS

16. Reduction of Algebraic Fractions.

Frequently the integral to be found, either before or after the process of substitution, takes the form of a Rational Algebraic Fraction. As a preparatory step to the integration of such a form, the fraction must always be resolved into Partial Fractions. We assume that the student is already familiar with this process; if this is not the case he is referred to the standard text-books on Algebra.

We shall proceed to show that if this is done the result can practically always be expressed in terms of powers, logs, and inverse tangents.

If the numerator of the given fraction is of higher order than the denominator, we should first divide out as far as we can. The quotient thus formed can only consist of powers of x and constants and is immediately integrable.

In what follows we shall suppose this to have been done and concern ourselves only with the remaining fractional part, in which the numerator will be at least one lower in order than the denominator.

The four possible types of term.

There will, in general, be four types of term which can occur:

(1) Terms like $\dfrac{A}{Bx + C}$.

(2) Terms like $\dfrac{A}{(Bx + C)^n}$.

(3) Terms like $\dfrac{ax+b}{(Ax^2+Bx+C)}$,

    where Ax^2+Bx+C has no real factors.

(4) Terms like $\dfrac{ax+b}{\{Ax^2+Bx+C\}^n}$.

We shall confine our attention to the first three types, as the last, which involves the presence in the denominator of a doubly repeated irreducible quadratic factor, is uncommon and outside the scope of this work. The reader desirous of knowing how to deal with this case is referred to Williamson's *Integral Calculus*, p. 53, §§ 43, etc. We proceed to show that the other types are immediately integrable:

(1) $\displaystyle\int \frac{A\,dx}{Bx+C}$ is seen to be $\dfrac{A}{B}\log(Bx+C)$.

(2) $\displaystyle\int \frac{A\,dx}{(Bx+C)^n}$ is seen to be $-\dfrac{A}{(n-1)\,B}\cdot\dfrac{1}{(Bx+C)^{n-1}}$.

Each of these may be verified by writing $Bx+C=Z$.

17. Integration of terms like $\dfrac{ax+b}{Ax^2+Bx+C}$.

In order to integrate terms of type (3) *we first introduce into the numerator a multiple of the differential of the denominator sufficient to remove all the x's*, making the numerical factors right afterwards.

Thus for $\dfrac{ax+b}{Ax^2+Bx+C}$ we put

$$\frac{a}{2A}\cdot\frac{2Ax+B}{Ax^2+Bx+C}+\left(b-\frac{aB}{2A}\right)\frac{1}{Ax^2+Bx+C}.$$

The integral of the term thus reduces to the sum of two integrals.

$$(a) \quad \frac{a}{2A} \int \frac{2Ax + B}{Ax^2 + Bx + C} dx,$$

and

$$(b) \quad \frac{2Ab - aB}{2A} \int \frac{dx}{Ax^2 + Bx + C}.$$

(a) is seen at once to be $\dfrac{a}{2A} \log (Ax^2 + 2Bx + C)$, (§ 6).

To deal with (b) we re-write the term as below

$$\frac{2Ab - aB}{2A} \int \frac{dx}{Ax^2 + Bx + C} = \frac{2Ab - aB}{2A^2} \int \frac{dx}{x^2 + \frac{B}{A} x + \frac{C}{A}},$$

$$= \frac{2Ab - aB}{2A^2} \int \frac{dx}{\left(x + \frac{B}{2A} \right)^2 + \frac{4AC - B^2}{4A^2}} dx.$$

Now since $Ax^2 + Bx + C$ has no real factors, $4AC - B^2$ is positive, i.e. the integral reduces to some integral of the form

$$K \int \frac{dx}{(x + M)^2 + N^2},$$

the result of which is seen to be $\dfrac{K}{N} \tan^{-1} \dfrac{x + M}{N}$.

$$Ex. \quad \int \frac{x + 5}{x^2 + 4x + 7} dx = \frac{1}{2} \int \frac{2x + 4}{x^2 + 4x + 7} dx + \int \frac{3}{x^2 + 4x + 7} dx$$

$$= \frac{1}{2} \log (x^2 + 4x + 7) + \int \frac{3}{(x + 2)^2 + 3} dx$$

$$= \frac{1}{2} \log (x^2 + 4x + 7) + \frac{3}{\sqrt{3}} \tan^{-1} \frac{x + 2}{\sqrt{3}} + C.$$

The reader should make no attempt to remember and quote these results as formulæ. It is far better to treat each example on its own merits. The articles above give the methods of attack and indicate the results likely to be found. The student with a knowledge of the use of imaginary quantities and de Moivre's Theorem will find that the case of irreducible factors can also be treated by the use of imaginary quantities, but this lies beyond the scope of the present volume.

18. Worked Examples.

(1) Find
$$\int \frac{dx}{x^2 - a^2}.$$

Let
$$\frac{1}{x^2 - a^2} = \frac{A}{x - a} + \frac{B}{x + a}.$$

Clearing of fractions we have $1 \equiv A(x + a) + B(x - a)$.

Since this is an identity we equate coefficients of powers of x.

$$\therefore \ A + B = 0,$$

$$A - B = \frac{1}{a},$$

whence
$$A = \frac{1}{2a}, \quad B = -\frac{1}{2a}.$$

$$\therefore \ \frac{1}{x^2 - a^2} \equiv \frac{1}{2a} \cdot \frac{1}{x - a} - \frac{1}{2a} \cdot \frac{1}{x + a}.$$

$$\therefore I = \frac{1}{2a} \int \frac{1}{x - a} \, dx - \frac{1}{2a} \int \frac{1}{x + a} \, dx$$

$$= \frac{1}{2a} \log (x - a) - \frac{1}{2a} \log (x + a) + C$$

$$= \frac{1}{2a} \log \frac{x - a}{x + a} + C.$$

Hence
$$\int \frac{dx}{x^2 - a^2} = \frac{1}{2a} \log \frac{x - a}{x + a} + C.$$

(2) $\int \dfrac{x}{(x^2 + 4x + 5)(x - 1)}\, dx.$

Let $\qquad \dfrac{x}{(x^2 + 4x + 5)(x - 1)} = \dfrac{A}{x - 1} + \dfrac{Bx + C}{x^2 + 4x + 5}.$

Proceeding in the ordinary way we find

$$A = \tfrac{1}{10}, \quad B = -\tfrac{1}{10}, \quad C = \tfrac{1}{2}.$$

Therefore

$$I = \frac{1}{10}\int \frac{dx}{x - 1} - \frac{1}{10}\int \frac{x - 5}{x^2 + 4x + 5}\, dx$$

$$= \frac{1}{10}\log(x - 1) - \frac{1}{20}\int \frac{2x + 4}{x^2 + 4x + 5}\, dx + \frac{7}{10}\int \frac{dx}{(x + 2)^2 + 1}$$

$$= \frac{1}{10}\log(x - 1) - \frac{1}{20}\log(x^2 + 4x + 5) + \frac{7}{10}\tan^{-1}(x + 2) + C.$$

(3) $\int \dfrac{x^3}{(x - 1)^2 (2x^2 + x + 1)}\, dx.$

$$\text{Exp.} = \frac{A}{x - 1} + \frac{B}{(x - 1)^2} + \frac{Cx + D}{2x^2 + x + 1}.$$

Clearing of fractions we get

$$x^3 \equiv A(x - 1)(2x^2 + x + 1) + B(2x^2 + x + 1) + (Cx + D)(x - 1)^2.$$

We get B by putting $x = 1$, whence $B = \tfrac{1}{4}$.

Equating the three highest coefficients we get

$$\left.\begin{array}{l} 1 = 2A + C \\ 0 = -A + 2B - 2C + D \\ 0 = B + C - 2D \end{array}\right\}.$$

Solving these equations

$$A = \tfrac{5}{8}, \quad B = \tfrac{1}{4}, \quad C = -\tfrac{2}{16}, \quad D = -\tfrac{3}{16}.$$

P.C.   3

$$\therefore I = \int \frac{5}{8} \cdot \frac{1}{x-1} \, dx + \int \frac{1}{4} \cdot \frac{1}{(x-1)^2} \, dx - \int \frac{2x+3}{16\,(2x^2+x+1)} \, dx$$

$$= \int \frac{5}{8} \cdot \frac{1}{x-1} \, dx + \int \frac{1}{4} \cdot \frac{1}{(x-1)^2} \, dx - \int \frac{1}{32} \cdot \frac{4x+1}{2x^2+x+1} \, dx$$

$$- \int \frac{5}{32} \cdot \frac{1}{2x^2+x+1} \, dx$$

$$= \frac{5}{8} \log(x-1) - \frac{1}{4} \cdot \frac{1}{x-1} - \frac{1}{32} \log(2x^2+x+1)$$

$$- \int \frac{5}{64} \cdot \frac{dx}{(x+\frac{1}{2})^2 + \frac{1}{4}}$$

$$= \frac{5}{8} \log(x-1) - \frac{1}{4} \cdot \frac{1}{x-1} - \frac{1}{32} \log(2x^2+x+1)$$

$$- \frac{5}{128} \tan^{-1}(2x+1) + C.$$

EXAMPLES IV

Integrate the following by the method of the last chapter:

(1) $\dfrac{1}{x^2+4x+3}$.   (2) $\dfrac{x}{x^2+2x+3}$.   (3) $\dfrac{x}{(4-5x)^2}$.

(4) $\dfrac{x}{x^2-5x+6}$.   (5) $\dfrac{2x^2+3x+4}{x^2+6x+10}$.   (6) $\dfrac{(x-1)^2}{x^2+2x+2}$.

*(7) $\dfrac{3}{x^3-3x^2+2}$.   (8) $\dfrac{x^2+1}{x^3+1}$.   (9) $\dfrac{1}{x^3+x^2+x}$.

Find:

(10) $\displaystyle\int_0^1 \frac{x^2+x+1}{x^2-x+1} \, dx$.   (11) $\displaystyle\int \frac{x^2\,dx}{x^4-x^2-12}$.

(12) $\displaystyle\int \frac{dx}{(x^2+a^2)\,(x^2+b^2)}$.   (13) $\displaystyle\int \frac{dx}{x^4+1}$.

(14) $\displaystyle\int \frac{\cos x \, dx}{16-9\sin^2 x}$.   (15) $\displaystyle\int_0^{\frac{\pi}{2}} \frac{\cos x \, dx}{(1+\sin x)\,(5+\sin x)}$.

(16) $\displaystyle\int_0^{\infty} \frac{x^2\,dx}{(x^2+a^2)\,(x^2+b^2)}$.

CHAPTER V

TRIGONOMETRICAL INTEGRALS
AND FORMULÆ OF REDUCTION

19. Trigonometrical Integrals. Odd power of sine or cosine.

We consider in this chapter various types of trigonometrical integrals frequently met with.

Any odd power of a sine or cosine, or any product of the form $\sin^a x \cos^b x$, where either of the indices a or b is an odd integer, can be evaluated at once by substitution. The method is clear from the following examples.

(a) $\displaystyle\int \sin^7 x\, dx = \int \sin^6 x \cdot \sin x\, dx.$

Let $\qquad z = \cos x, \ -dz = \sin x\, dx.$

Then $I = -\displaystyle\int (1 - z^2)^3\, dz$

$\qquad = \displaystyle\int (z^6 - 3z^4 + 3z^2 - 1)\, dz$

$\qquad = \dfrac{z^7}{7} - \dfrac{3z^5}{5} + \dfrac{3z^3}{3} - z + C$

$\qquad = \tfrac{1}{7}\cos^7 x - \tfrac{3}{5}\cos^5 x + \cos^3 x - \cos x + C.$

(b) Or again, to integrate $\dfrac{\sin^4 x}{\cos^5 x}$,

$$\int \frac{\sin^4 x}{\cos^5 x} = \int \frac{\sin^4 x}{\cos^6 x} \cdot \cos x\, dx.$$

Let $\qquad z = \sin x, \ dz = \cos x\, dx.$

Then I becomes $\displaystyle\int \frac{z^4}{(1 - z^2)^3}\, dz.$

Now this is an algebraic fraction and can clearly be evaluated by the methods of a preceding chapter. We need not continue the solution, which is laborious; it is at any rate clear that the integral is no longer trigonometrical.

We see that the general principle is to choose the substitutions so that the function of which there is an odd power is the differential and the integral then becomes algebraic in form. The case when both indices are even presents a greater difficulty and is left till later in the chapter.

20. Integrals of the form

$$\int \frac{dx}{a + b \sin x}, \quad \int \frac{dx}{a + b \cos x}.$$

A second large class of trigonometrical integrals are of the form

$$\int \frac{dx}{a + b \cos x} \quad \text{or} \quad \int \frac{dx}{a + b \sin x}.$$

These, in common with many other trigonometrical integrals, are best dealt with by making the tangent of half the angle the subject of integration.

If $t = \tan \dfrac{x}{2}$, we know that

$$\sin x = \frac{2t}{1 + t^2}, \quad \cos x = \frac{1 - t^2}{1 + t^2}.$$

Also
$$dt = \frac{1}{2} \sec^2 \frac{x}{2} \, dx,$$

i.e.
$$\frac{2dt}{1 + t^2} = dx.$$

The details of this transformation are well worth committing to memory. The use of them will cause the integrals we are considering, and many other trigonometrical integrals, to take an algebraic form, and integration can then be completed by the methods of Ch. IV.

21. Worked Examples.

(1) Find $\quad \int \sec x\, dx \quad$ and $\quad \int \operatorname{cosec} x\, dx.$

Let $\qquad t = \tan \dfrac{x}{2}, \quad dx = \dfrac{2dt}{1+t^2}.$

$$\therefore \int \operatorname{cosec} x\, dx = \int \frac{1+t^2}{2t} \cdot \frac{2dt}{1+t^2} = \int \frac{1}{t} \cdot dt$$

$$= \log t = \log \tan \frac{x}{2} + C.$$

Similarly $\displaystyle \int \sec x\, dx = \int \frac{1+t^2}{1-t^2} \cdot \frac{2dt}{1+t^2}$

$$= \int \frac{2}{1-t^2}\, dt$$

$$= \int \left(\frac{1}{1-t} + \frac{1}{1+t} \right) dt$$

$$= \log \frac{1 + \tan \dfrac{x}{2}}{1 - \tan \dfrac{x}{2}} + C,$$

or $\qquad\qquad \log \tan \left(\dfrac{\pi}{4} + \dfrac{x}{2} \right) + C.$

We thus have two new standard forms

$$\int \sec x\, dx = \log \tan \left(\frac{\pi}{4} + \frac{x}{2} \right) + C,$$

$$\int \operatorname{cosec} x\, dx = \log \tan \frac{x}{2} + C.$$

(2) $\displaystyle\int \frac{dx}{5 + 3\sin x}$.

Making the same substitutions as above we get

$$
\begin{aligned}
I &= \int \frac{\dfrac{2dt}{1+t^2}}{5 + \dfrac{6t}{1+t^2}} \\
&= \int \frac{2dt}{5 + 6t + 5t^2} \\
&= \frac{2}{5}\int \frac{dt}{t^2 + \frac{6}{5}t + 1} \\
&= \frac{2}{5}\int \frac{dt}{(t + \frac{3}{5})^2 + (\frac{4}{5})^2} \\
&= \frac{2}{5}\cdot\frac{5}{4}\tan^{-1}\frac{t + \frac{3}{5}}{\frac{4}{5}} + C \\
&= \tfrac{1}{2}\tan^{-1}\frac{5\tan\dfrac{x}{2} + 3}{4} + C.
\end{aligned}
$$

The student will observe that all integrals of this type eventually become of the form $\displaystyle\int \frac{1}{z^2 + a^2}\,dz$ or $\displaystyle\int \frac{1}{z^2 - a^2}\,dz$ and hence the functions appearing in the answer are usually likely to be inverse tangents or logarithms. The same method and remarks apply to any integral of the form $\displaystyle\int \frac{dx}{a + b\cos x + c\sin x}$. But the student will best make himself familiar with the forms that can occur by examples.

EXAMPLES V A

Integrate the following:

(1) $\sin^3 x$.   (2) $\sin^5 x$.   (3) $\cos^9 x$.   (4) $\sin^2 x \cos^5 x$.

(5) $\dfrac{\cos^5 x}{\sin^4 x}$.   (6) $\dfrac{\cos^2 x}{\sin^3 x}$.   (7) $\sin 2x \cos^4 x$.

(8) $\sin 3x \cos^4 x$.   (9) $\dfrac{1}{3 + 5 \cos x}$.   (10) $\dfrac{1}{3 - 5 \sin x}$.

(11) $\dfrac{1}{\sin x + \cos x}$.   (12) $\dfrac{1}{1 + \sin x + \cos x}$.   (13) $\dfrac{1}{2 + \sin x + \cos x}$.

(14) $\dfrac{\sec x}{a + b \tan x}$.

*(15) Show that

$$\int \frac{\sin x}{\cos 2x} \, dx = \frac{1}{2\sqrt{2}} \log \cot\left(\frac{\pi}{8} + \frac{x}{2}\right) \cot\left(\frac{\pi}{8} - \frac{x}{2}\right) + C.$$

*(16) Prove that

$$\int \frac{\sin \theta \, d\theta}{3 \cos \theta + 4 \sin \theta} = \frac{4}{25} \theta + \frac{3}{25} \log (3 \cos \theta + 4 \sin \theta) + C.$$

Find

*(17) $\displaystyle\int_0^{\frac{\pi}{2}} \frac{dx}{1 - \cos a \sin x}$.   (18) $\displaystyle\int_0^{\frac{\pi}{2}} \frac{1}{1 - k^2 \sin^2 x}$.

(19) $\displaystyle\int \frac{dx}{\cos a + \cos x}$.

(20) Prove that, if $a < 1$,

$$\int_0^{\pi} \frac{dx}{1 - 2a \cos x + a^2} = \frac{\pi}{1 - a^2}.$$

22. Integrals of even order.

In a previous section we postponed the discussion of trigonometrical integrals of even order. There is no general rule for these integrals which can be readily obtained from first principles.

For small values of the indices the integrals may often be evaluated by the aid of multiple angles.

For example $\cos^2 \theta = \dfrac{1 + \cos 2\theta}{2}$.

Hence $\displaystyle\int \cos^2 \theta \, d\theta = \frac{\theta}{2} + \frac{\sin 2\theta}{4} + C.$

Or again $\displaystyle\int \sin^2 \theta \cos^2 \theta\, d\theta = \int \frac{1}{4} \sin^2 2\theta\, d\theta$

$$= \int \left[\frac{1}{8} - \frac{\cos 4\theta}{8}\right] d\theta$$

$$= \frac{\theta}{8} - \frac{\sin 4\theta}{32} + C.$$

But although this can always theoretically be done, it is clear that the method is too laborious to be of any practical use.

In certain other cases the substitution $x = \tan\theta$ may be employed.

For example, consider $\displaystyle\int \frac{\sin^2 \theta}{\cos^6 \theta}\, d\theta$.

Put $x = \tan\theta, \quad dx = \sec^2\theta\, d\theta,$

i.e. $\dfrac{dx}{1+x^2} = d\theta.$

Also $\sin\theta = \dfrac{x}{\sqrt{1+x^2}} \quad \text{and} \quad \cos\theta = \dfrac{1}{\sqrt{1+x^2}}.$

Hence $\displaystyle I = \int \frac{x^2}{1+x^2} \cdot \frac{(1+x^2)^2}{1} \cdot dx$

$$= \int x^2 (1+x^2)\, dx$$

$$= \int x^2 + x^4 dx$$

$$= \frac{x^3}{3} + \frac{x^5}{5} + C.$$

$\therefore \;\; I = \tfrac{1}{3}\tan^3\theta + \tfrac{1}{5}\tan^5\theta + C.$

Here again it is clear that the scope of the method is limited.

23. Reduction Formulæ.

The discussion of the general case brings us at once to a mention of Reduction Formulæ. Space and the professed scope of this volume will not allow us to deal fully with this subject—but we hope that the student will at least acquire a slight familiarity with the chief forms.

We have already seen how many integrals, themselves apparently insoluble, can be connected in some way or other with other integrals more readily found. In a manner of speaking any such device for connecting one integral with another is a "reduction formula"—for instance, the method of Integration by Parts might be so termed. But the term "reduction formula" is usually restricted to methods of connecting one integral with another of the same type but of lower order, a process which may have to be repeated a number of times before we arrive at an integral easily evaluated. This method is particularly of use in dealing with such forms as $\int \sin^m x\,dx, \int \cos^m x\,dx,$ $\int \cos^n x \sin^m x\,dx$. The actual reduction formulæ are usually found by trial, as in the following examples.

24. To find reduction formulæ for

$$\int \sin^n x\,dx \quad \text{and} \quad \int \cos^n x\,dx.$$

Consider first $\qquad \int \sin^n x\,dx.$

We endeavour to connect this integral with $\int \sin^{n-2} x\,dx$.

Let $\qquad P = \sin^{n-1} x \cos x.$

Then $\dfrac{dP}{dx} = (n-1)\sin^{n-2}x\cos^2 x - \sin^{n-1}x\,.\,\sin x$

$\qquad\qquad = (n-1)\sin^{n-2}x\,(1-\sin^2 x) - \sin^n x$

$\qquad\qquad = (n-1)\sin^{n-2}x - n\sin^n x.$

Hence integrating both sides we get

$$P = (n-1)\int \sin^{n-2}x\,dx - n\int\sin^n x\,dx,$$

or $\displaystyle\int\sin^n x\,dx = -\dfrac{1}{n}\{\sin^{n-1}x\cos x\} + \dfrac{n-1}{n}\int\sin^{n-2}x\,dx.$

Similarly, putting $n-2$ for n,

$$\int\sin^{n-2}x\,dx = -\dfrac{1}{n-2}\{\sin^{n-3}x\cos x\} + \dfrac{n-3}{n-2}\int\sin^{n-4}x\,dx,$$

and substituting we can connect

$$\int\sin^n x\,dx \quad\text{with}\quad \int\sin^{n-4}x\,dx.$$

Proceeding in this way we see that $\displaystyle\int\sin^n x\,dx$ can finally be connected *either* with $\displaystyle\int\sin x\,dx$ (if n is an odd integer) *or* with $\displaystyle\int 1\,dx$ (if n is an even integer), and thus the integral can finally be evaluated.

$\displaystyle\int\cos^n x\,dx$ can be found in exactly the same way by choosing $P = \cos^{n-1}x\sin x.$

If n is a negative integer, we can instead connect

$$\int\sin^n x\,dx \quad\text{with}\quad \int\sin^{n+2}x\,dx$$

by choosing $P = \sin^{n+1}x\cos x$ and following the same method; and likewise for $\displaystyle\int\cos^n x\,dx$ when n is negative we choose $P = \cos^{n+1}x\sin x.$

25. Two important definite integrals.

These methods are particularly useful in dealing with certain definite integrals, arising in problems of mensuration, etc. The commonest integrals of this type are those in which the limits are $\frac{\pi}{2}$ and zero or multiples of these limits.

We proceed to show how $\int_0^{\frac{\pi}{2}} \sin^n x\,dx$ and $\int_0^{\frac{\pi}{2}} \cos^n x\,dx$ can be written down at once with the aid of these methods.

We have seen that

$$\int \sin^n x\,dx = -\frac{1}{n}\left\{\sin^{n-1}x \cos x\right\} + \frac{n-1}{n}\int \sin^{n-2}x\,dx,$$

$$\therefore \int_0^{\frac{\pi}{2}} \sin^n x\,dx = \frac{n-1}{n}\int_0^{\frac{\pi}{2}} \sin^{n-2}x\,dx,$$

for the quantity $-\frac{1}{n}\left\{\sin^{n-1}x \cos x\right\}$ vanishes for both limits.

Denote $\int_0^{\frac{\pi}{2}} \sin^n x\,dx$, which is clearly a function of n, by u_n. Then we have successively

$$\left.\begin{array}{l} u_n = \dfrac{n-1}{n}\,u_{n-2} \\[2mm] u_{n-2} = \dfrac{n-3}{n-2}\,u_{n-4} \\[2mm] \text{etc.} \end{array}\right\} \quad \ldots\ldots\ldots\ldots(A).$$

Two cases naturally arise: (1) n odd, (2) n even.

Case I. If n is odd, the last of the equations (A) is

$$u_3 = \tfrac{2}{3}u_1.$$

Hence eliminating all the intermediate functions we get

$$u_n = \frac{n-1}{n} \cdot \frac{n-3}{n-2} \cdots \cdot \frac{4}{5} \cdot \frac{2}{3} u_1.$$

But
$$u_1 = \int_0^{\frac{\pi}{2}} \sin x\, dx = \left[-\cos x\right]_0^{\frac{\pi}{2}} = 1.$$

$$\therefore \text{ Finally } \int_0^{\frac{\pi}{2}} \sin^n x\, dx \{n \text{ odd}\} \equiv \frac{n-1}{n} \cdot \frac{n-3}{n-2} \cdots \cdot \frac{2}{3}.$$

Case II. If, however, n is even, the last of the relations (A) will be $u_2 = \frac{1}{2}u_0$.

Hence eliminating all intermediate functions we now get

$$u_n = \frac{n-1}{n} \cdot \frac{n-3}{n-2} \cdots \cdot \frac{1}{2} u_0.$$

But
$$u_0 = \int_0^{\frac{\pi}{2}} \sin^0 x\, dx = \int_0^{\frac{\pi}{2}} 1\, dx = \frac{\pi}{2}.$$

$\therefore$ Finally

$$\int_0^{\frac{\pi}{2}} \sin^n x\, dx \{n \text{ even}\} \equiv \frac{n-1}{n} \cdot \frac{n-3}{n-2} \cdots \cdot \frac{1}{2} \cdot \frac{\pi}{2}.$$

The reader should notice carefully how these results are written down, as it is often useful to be able to quote them.

We place the index n in the first denominator, writing above it $n-1$ as the first numerator. We proceed thus, writing first the denominator and then the numerator till we come to the figure 2. If 2 is a numerator we leave the result as it stands, but if 2 is a denominator we add another factor $\frac{\pi}{2}$. The reader should make himself proficient in this by a few examples of his own choice.

26. $\displaystyle\int_0^{\frac{\pi}{2}} \cos^n x\, dx = \int_0^{\frac{\pi}{2}} \sin^n x\, dx.$

It is quite easy to find $\int \cos^n x\, dx$ in the same manner, but it is more instructive to prove that

$$\int_0^{\frac{\pi}{2}} \sin^n x\, dx \quad \text{and} \quad \int_0^{\frac{\pi}{2}} \cos^n x\, dx$$

are equal.

For let $\qquad\qquad x = \dfrac{\pi}{2} - y,$

i.e. $\qquad\qquad\qquad dx = - dy,$

and when the limits for x are $\dfrac{\pi}{2}$ and 0, those for y are 0 and $\dfrac{\pi}{2}$.

Hence $\qquad \displaystyle\int_0^{\frac{\pi}{2}} \sin^n x\, dx = \int_{\frac{\pi}{2}}^0 \cos^n y\,(-dy)$

$$= -\int_{\frac{\pi}{2}}^0 \cos^n y\, dy.$$

But clearly to change the order of the limits changes the sign of the integrals.

$$\therefore \int_0^{\frac{\pi}{2}} \sin^n x\, dx = +\int_0^{\frac{\pi}{2}} \cos^n y\, dy.$$

Hence the results we have just proved for

$$\int_0^{\frac{\pi}{2}} \sin^n x\, dx \quad \text{apply equally to} \quad \int_0^{\frac{\pi}{2}} \cos^n x\, dx.$$

27. Other Limits.

But care must be exercised in applying any of these results when the limits are other than $\frac{\pi}{2}$ and 0; for instance, π and 0.

Thus while it is correct to state that

$$\int_0^\pi \sin^n x\,dx = 2\int_0^{\frac{\pi}{2}} \sin^n x\,dx,$$

the same result is only true for cosines if n is even; if n is odd the value of $\int_0^\pi \cos^n x\,dx$ being zero.

This is obvious from the fact that the sine repeats its positive values in the second quadrant, whilst the cosine changes sign; or may be proved rigidly as follows:

Now $\displaystyle\int_0^\pi \cos^n x\,dx = \int_{\frac{\pi}{2}}^\pi \cos^n x\,dx + \int_0^{\frac{\pi}{2}} \cos^n x\,dx.$

Let $y = \pi - x,$

$$dy = -\,dx.$$

Also if limits for x are π and $\frac{\pi}{2}$, those for y are $0, \frac{\pi}{2}$.

$$\therefore \int_{\frac{\pi}{2}}^\pi \cos^n x\,dx = \int_{\frac{\pi}{2}}^0 \{-\cos y\}^n \cdot -dy$$

$$= \int_0^{\frac{\pi}{2}} (-1)^n \cos^n y\,dy$$

$$= +\int_0^{\frac{\pi}{2}} \cos^n y\,dy \quad \text{if } n \text{ is even,}$$

but $-\int_0^{\frac{\pi}{2}} \cos^n y\,dy \quad \text{if } n \text{ is odd,}$

and hence the whole integral

$$= 0 \quad \text{if } n \text{ is odd},$$

but $$= 2 \int_0^{\frac{\pi}{2}} \cos^n x \, dx \quad \text{if } n \text{ is even.}$$

The only safe rule for the beginner is to treat every example on its own merits, and, wherever the least doubt arises, make some substitution such as the above to verify his conclusions.

28. The form $\int \sin^p x \cos^q x \, dx$.

The method of successive reductions can also be applied to any integral of the form $\int \sin^p x \cos^q x \, dx$. This can be connected with a number of other integrals of parallel form, the methods being analogous in each case.

We will content ourselves here with exhibiting the connection between

$$\int \sin^p x \cos^q x \, dx \quad \text{and} \quad \int \sin^p x \cos^{q-2} x \, dx.$$

The other reductions will be found in the set of examples at the end of the chapter.

As before we take one more sine and one less cosine, i.e. we start with $P = \sin^{p+1} x \cos^{q-1} x$.

$$\therefore \frac{dP}{dx} = (p+1) \sin^p x \cos^q x - (q-1) \sin^{p+2} x \cos^{q-2} x$$

$$= (p+1) \sin^p x \cos^q x$$
$$\qquad - (q-1) \sin^p x (1 - \cos^2 x) \cos^{q-2} x$$

$$= (p+q) \sin^p x \cos^q x - (q-1) \sin^p x \cos^{q-2} x.$$

$\therefore$ Integrating both sides we have

$$P = (p+q) \int \sin^p x \cos^q x \, dx - (q-1) \int \sin^p x \cos^{q-2} x \, dx.$$

Hence

$$\int \sin^p x \cos^q x\, dx = -\frac{\sin^{p+1} x \cos^{q-1} x}{p+q}$$
$$+ \frac{q-1}{p+q} \int \sin^p x \cos^{q-2} x\, dx.$$

The same process may be applied to the latter integral and we can proceed thus, reducing the order of the integral step by step until either all the cosines have disappeared or only one cosine is left.

In the first case we can go on to evaluate $\int \sin^p x\, dx$ by the methods of § 24; in the second case the integral may be solved by the substitution $y = \sin x$.

In practice only the first case is likely to occur, as when either p or q is odd the integral can be dealt with much more easily by substitution, as explained in the earlier part of this chapter.

The connection with $\int \sin^{p-2} x \cos^q x\, dx$ can be effected in an exactly similar manner.

29. The definite integral $\int_0^{\frac{\pi}{2}} \sin^p x \cos^q x\, dx$.

As before, a case which occurs very frequently is that in which p and q are both positive integers, and the limits are $\frac{\pi}{2}$ and 0.

In this case it is easy to see that the reduction formula is simply

$$\int_0^{\frac{\pi}{2}} \sin^p x \cos^q x\, dx = \frac{q-1}{p+q} \int_0^{\frac{\pi}{2}} \sin^p x \cos^{q-2} x\, dx.$$

Again

$$\int_0^{\frac{\pi}{2}} \sin^p x \cos^{q-2} x\, dx = \frac{q-3}{p+q-2} \int_0^{\frac{\pi}{2}} \sin^p x \cos^{q-4} x\, dx,$$

and so on.

Assuming p and q even we get finally

$$\int_0^{\frac{\pi}{2}} \sin^p x \cos^2 x\, dx = \frac{1}{p+2} \int_0^{\frac{\pi}{2}} \sin^p x\, dx$$

$$= \frac{1}{p+2} \cdot \frac{p-1}{p} \cdot \frac{p-3}{p-2} \cdots \cdot \frac{1}{2} \cdot \frac{\pi}{2}.$$

Hence finally, eliminating all intermediate integrals, we have

$$\int_0^{\frac{\pi}{2}} \sin^p x \cos^q x\, dx \ (p \text{ and } q \text{ both even})$$

$$\equiv \frac{(p-1)(p-3)\ldots\ldots 3 \cdot 1 \cdot (q-1)(q-3)\ldots\ldots 3 \cdot 1}{(p+q)(p+q-2)\ldots\ldots 4 \cdot 2} \cdot \frac{\pi}{2}.$$

It is evident from the symmetry of the result that

$$\int_0^{\frac{\pi}{2}} \sin^p x \cos^q x\, dx = \int_0^{\frac{\pi}{2}} \sin^q x \cos^p x\, dx,$$

when p and q are both even, a result which can easily be proved directly.

30. In the case where either p or q is negative we connect the integral with one of apparently higher order, e.g. if p is negative we connect

$$\int \sin^p x \cos^q x\, dx \ \text{ with } \ \int \sin^{p+2} x \cos^q x\, dx,$$

by taking $P = \sin^{p+1} x \cos^{q+1} x$ and differentiating, the steps being exactly as before. By this method the index of $\sin x$ is successively raised by 2 at each reduction until it is finally either zero or unity, in either of which cases integration can be effected by methods mentioned in various parts of this chapter.

In fact, by correctly choosing P, the integral may be connected with any of the integrals

$$\int \sin^{p \pm 2} x \cos^q x \, dx,$$

$$\int \sin^p x \cos^{q \pm 2} x \, dx,$$

$$\int \sin^{p \pm 2} x \cos^{q \pm 2} x \, dx.$$

The reader is left to find the details of these reductions for himself.

31. The definite integrals of §25 lead to a curious result, first discovered by Wallis.

Since $\quad \sin x < 1, \quad \sin^{n-1} x > \sin^n x > \sin^{n+1} x.$

$$\therefore \int_0^{\frac{\pi}{2}} \sin^{n-1} x \, dx > \int_0^{\frac{\pi}{2}} \sin^n x \, dx > \int_0^{\frac{\pi}{2}} \sin^{n+1} x \, dx.$$

Let n be even and positive.

Then

$$\frac{n-2}{n-1} \cdot \frac{n-4}{n-3} \cdots \cdot \frac{2}{3} > \frac{n-1}{n} \cdot \frac{n-3}{n-2} \cdots \cdot \frac{1}{2} \cdot \frac{\pi}{2}$$

$$> \frac{n}{n+1} \cdot \frac{n-2}{n-1} \cdots \cdot \frac{2}{3}.$$

Hence $\quad \dfrac{\pi}{2} > \dfrac{2^2 \cdot 4^2 \cdots (n-2)^2 \cdot n^2}{1 \cdot 3^2 \cdot 5^2 \cdots (n-1)^2 (n+1)},$

but $\quad < \dfrac{2^2 \cdot 4^2 \cdots (n-2)^2 \cdot n}{1^2 \cdot 3^2 \cdot 5^2 \cdots (n-1)^2},$

and since the ratio of these fractions is $\dfrac{n}{n+1}$, which is ultimately unity, the result is sometimes stated in the form

$$\frac{\pi}{2} = \operatorname*{Lt}_{n \to \infty} \frac{2^2 \cdot 4^2 \cdot \ldots \cdot (n-2)^2 \cdot n}{1 \cdot 3^2 \cdot 5^2 \cdot \ldots \cdot (n-1)^2}.$$

These are known as Wallis' Formulæ.

EXAMPLES V B

Integrate the following:

(1) $\sin^4 x$.   (2) $\cos^6 x$.   (3) $\sin^2 x \cos^2 x$.   (4) $\sin^4 x \cos^2 x$.

(5) $\dfrac{1}{\sin^2 x \cos^2 x}$.   (6) $\dfrac{1}{\sin^3 x}$.   (7) $\dfrac{1}{\sin x \cos^2 x}$.

(8) Find a reduction formula for $\int \sin^n x\, dx$ when n is negative.

(9) Connect $\int \sin^p x \cos^q x\, dx$ with

   (a) $\int \sin^{p+2} x \cos^{q-2} x\, dx$,   ($b$) $\int \sin^{p-2} x \cos^{q+2} x\, dx$.

Write down the values of the following:

(10) $\displaystyle\int_0^{\frac{\pi}{2}} \sin^8 \theta\, d\theta$, $\displaystyle\int_0^{\pi} \sin^8 \theta\, d\theta$, $\displaystyle\int_0^{\pi} \cos^8 \theta\, d\theta$.

(11) $\displaystyle\int_0^{\frac{\pi}{2}} \cos^6 \theta\, d\theta$, $\displaystyle\int_0^{\frac{\pi}{2}} \cos^7 \theta\, d\theta$, $\displaystyle\int_0^{\pi} \cos^7 \theta\, d\theta$, $\displaystyle\int_{-\pi}^{\pi} \cos^7 \theta\, d\theta$.

(12) $\displaystyle\int_0^{\frac{\pi}{2}} \sin^4 \theta \cos^4 \theta\, d\theta$, $\displaystyle\int_0^{\pi} \sin^4 \theta \cos^4 \theta\, d\theta$.

*(13) $\displaystyle\int_0^{\frac{\pi}{2}} (2 \sin^5 x - 3 \sin^7 x) \cos^2 x\, dx$.

(14) Prove the formula

$$\int \sec^{2n+1} \phi\, d\phi = \frac{1}{2n} \tan \phi \sec^{2n-1} \phi + \frac{2n-1}{2n} \int \sec^{2n-1} \phi\, d\phi.$$

(15) Integrate $\sec^3 \phi$ and $\sec^5 \phi$.

(16) Prove

$$\int x^m (\log x)^p \, dx = \frac{x^{m+1} (\log x)^p}{m+1} - \frac{p}{m+1} \int x^m (\log x)^{p-1} \, dx$$

and integrate $x^4 (\log x)^2$.

(17) Show that $\quad \int_0^a x^2 (a^2 - x^2)^{\frac{3}{2}} = \frac{\pi a^6}{32}$.

(18) Find a reduction formula for

$$\int x^m (a^2 - x^2)^{\frac{p}{2}} \, dx.$$

(19) Show that by taking $P = x^{m+\frac{1}{2}} (2a - x)^{\frac{3}{2}}$ we can find a reduction formula for $\int x^m \sqrt{2ax - x^2} \, dx$.

(20) Calculate $\quad \int_0^{2a} x^4 \sqrt{2ax - x^2} \, dx.$

CHAPTER VI

THE FUNDAMENTAL "INVERSION THEOREM." APPLICATION OF CALCULUS TO FINDING OF AREAS

32. The two meanings of integration.

Up to the present we have defined the process of integration as being the reversal of that of differentiation and it is presumed that, by this time, the student has some idea of the main methods which can be employed in evaluating an integral from this point of view. Looked at in this light, integration is merely an artificial mathematical process, and, did it not possess another meaning, would quickly cease to be of any interest or importance. It is, however, quite reasonable to expect that this mathematical process will have some definitely physical meaning, for the process of which it is the reverse (viz. differentiation) is known to be the method of* finding the "rate of increase" of the given function. We will now proceed to show that the process of integration, or at least that of definite integration, is equivalent to that of finding the sum of a certain series of small quantities.

---

* The student should observe that there is a fundamental theorem of the same character in Differential, viz. that the result of the purely analytical process of finding $\underset{h \to 0}{\mathrm{Lt}} \dfrac{\phi(x+h) - \phi(x)}{h}$ is really the same as that of finding the "rate of increase" of $\phi(x)$. The proof of this theorem being easier than that of the Integral Theorem, we are rather apt to lose sight of its existence.

33. The Fundamental Theorem; Proof by consideration of an Area.

A rough idea of the theorem can be gained from the consideration of the problem of finding the area of the figure bounded by a given curve, two given ordinates, and the axis of x. In Integral Calculus this area is often referred to as the area of (or under) the curve between the given limits.

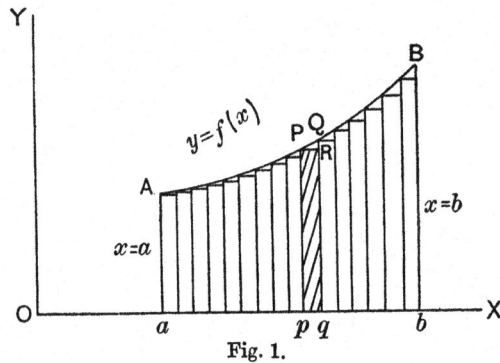

Fig. 1.

Suppose the equation of the bounding curve to be (in ordinary rectangular Cartesians) $y = f(x)$ and the bounding ordinates to be the lines $x = a$, $x = b$.

Divide the base into n equal divisions of width h and complete rectangles as in Fig. 1.

Obviously the sum of the areas of these rectangles constitutes an approximation to the area under the curve and this approximation can be made closer and closer by increasing the number of rectangles (and, of course, decreasing their width h). We may say, in fact, that the area under the curve is the limit of the sum of the areas of an indefinitely large number of rectangles of very small width, formed as in the figure.

Further, we can form an expression for the area of any such rectangle. For if the point of it which rests on the curve be (x_1, y_1) its area is $y_1 h$, i.e. $h \cdot f(x_1)$.

Or, in the limit, we may write this $f(x_1) \, \delta x$, since h is by definition the increase in x in passing from one rectangle to the next.

Hence the whole area is $\Sigma f(x) \, \delta x$, the summation being taken over all possible values of x lying between the limits $x = b$ and $x = a$.

But we may also look at the matter in a slightly different way. Suppose the area already reckoned up to a point $P(x, y)$ to be called $A(x)$, this notation being adopted to remind ourselves that A is a function of x.

Then in passing from P to an adjacent point Q we increase the area by an amount which we naturally denote by $\delta A(x)$. But the actual increase is the portion $PQqp$, which, as we have said, is closely approximated to by the rectangle whose area is $y \, \delta x$.

We have then $\qquad \delta A(x) = y \, \delta x$, approximately,

i.e. $$\frac{\delta A(x)}{\delta x} = y.$$

Or, in the limit, when the number of divisions is indefinitely increased,

$$\frac{d A(x)}{dx} = y.$$

Hence according to our definition of integration

$$A(x) = \int y \, dx + C = \int f(x) \, dx + C.$$

Now assume for the moment that $\int f(x) \, dx$ can be evaluated by the previous methods and that the result is $F(x)$.

Then $$A(x) = F(x) + C.$$

The constant C must clearly be chosen in such a way that when P and A coincide the result is zero, i.e.

$$C = -F(a).$$

Hence the area under the curve up to a point whose abscissa is x is $F(x) - F(a)$.

Hence the area required up to a point whose abscissa is b is $F(b) - F(a)$.

But this would be exactly the value of the definite integral $\int_a^b f(x)\,dx$, which is thus a second expression for the area under the curve.

These two expressions must, of course, be equal.

34. As an additional verification we can calculate the "apparent error" in this method.

The areas which are apparently neglected in Fig. 1 are a number of nearly triangular portions of which PQR is typical. Each of these "triangles" stands on an equal base and the sum of their respective heights is equal to the distance between the ordinates of A and B.

Hence the sum of all such "triangles" is $\frac{1}{2}h\,\{Bb \sim Aa\}$.

Now $(Bb \sim Aa)$ is definitely finite and hence the apparent error is a small quantity which $\rightarrow 0$ as $h \rightarrow 0$.

It is not in fact as large as the "mean" strip, for this would be

$$\tfrac{1}{2}h\,\{Bb + Aa\}.$$

This argument does not break down even in the case of Fig. 1 a.

For down to the turning point T the errors are of excess, but from T to B of defect. The errors must therefore be reckoned of opposite sign and we are still led to the same result, $\frac{1}{2}h\{Bb \sim Aa\}$, in whatever manner the figure is drawn. The student can verify this by a figure with two turning points between A and B.

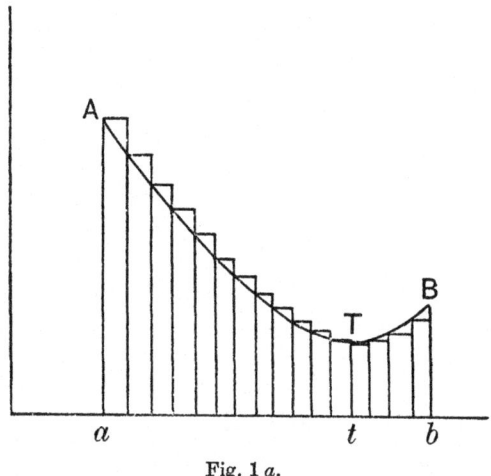

Fig. 1 a.

To give a purely analytical proof of this theorem is a matter of great intricacy and hardly required here. For further information the student is referred to Goursat's *Cours d'Analyse* and Hardy's *Pure Mathematics*. The proof given is, however, valid for any function that is capable of being represented by a continuous graph.

35. The Fundamental Theorem (*continued*).

We have thus arrived at the Fundamental Theorem of the Integral Calculus, which can be applied to integrals of any kind.

The operation $\int_a^b \phi(x)\, dx$ represents the summation of all possible expressions $\phi(a + rh) \cdot h$, where r is any integer, $h = \dfrac{b-a}{n}$ and $n \to \infty$.

Or putting the same statement in the form of a rough rule:

$\int_a^b \phi(x)\, dx$ is the sum of all expressions like $\phi(x)\, dx$ between $x = b$ and $x = a$.

The reader should notice that the method of proof given above implies certain restrictions on the nature of $f(x)$. It assumes, for example, that $f(x)$ is continuous; i.e. that the curve passes from A to B without any abrupt changes in the value of y, and further that no point exists between the end points for which $f(x)$ becomes infinite (though as a matter of fact a value of x for which $f(x) \to \infty$ may be one of the limits itself in certain cases).

This theorem, which seems to have been discovered independently by Newton and Leibnitz, enables us to write down at once the results of summations of this character, many of which are impossible to effect by ordinary algebraical methods of summation. The reader would do well to spend some time and trouble over the proof of this theorem before proceeding further. A thorough understanding of its principles is absolutely necessary to him, if he is to use the Calculus freely.

36. Areas of Curves.

It follows at once from the above work that areas included between any given curve and the axis of x may be calculated from the formula $\int y\, dx$; y being substituted in terms of x from the equation to the curve and the proper limits taken.

Similarly areas between the given curve and the axis of y may be calculated in the same manner in terms of y from the formula $\int x\,dy$.

The whole area of closed curves may be evaluated by adding or subtracting such areas, as may be seen from the diagrams.

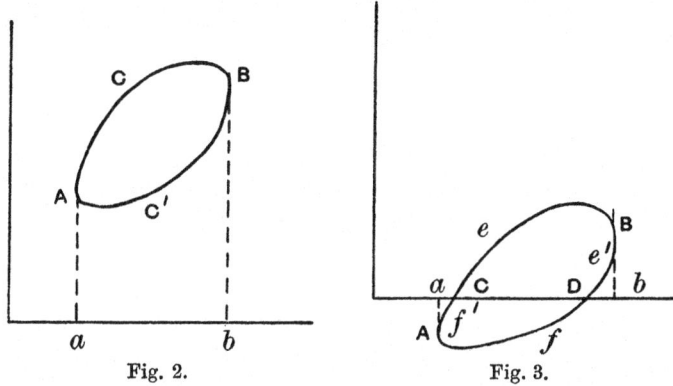

Fig. 2.

Fig. 3.

Fig. 2. Area of curve = area $ACBba$ – area $AC'Bba$.

Fig. 3. Area = $CeBbC - De'BbD + DfAaD - Cf'AaC$.

The reader can easily construct other diagrams for himself.

He will also notice that care needs to be exercised in the case of a curve which crosses the axis of x.

If, for example, he were performing the process suggested for Fig. 3, he would find that the portions $DfAaD$, $Cf'AaC$ would both come out with a negative sign (which he would have to disregard in performing the final additions and subtractions). This is easily seen to be the case, since the element $y\,dx$ changes sign at C and D.

In fact, for a portion of a curve crossing the axis between the limits, the ordinary formula gives the difference of the areas formed and not their sum. The only safe rule is to draw the figure and perform the integrations separately.

In the case of symmetrical curves it is often only necessary to find the area of half or a quadrant of the curve. For example, in finding the area of the ellipse $\dfrac{x^2}{a^2} + \dfrac{y^2}{b^2} = 1$, we need find only the area of the positive quadrant (included above the axis of x between $x = a$, $x = 0$) and multiply the result by 4.

37. Parametric Notations.

Many curves, in particular curves of the cycloid family, have their equations most simply expressed in parametric coordinates, i.e. x and y are both given in terms of some third variable t.

The expressions for the area then become $\displaystyle\int y\,\frac{dx}{dt}\,.\,dt$ and $\displaystyle\int x\,\frac{dy}{dt}\,.\,dt$, the proper limits being substituted for t and the whole question worked in terms of t.

Remember that when a curve is given in this form **it is almost always a mistake to attempt to eliminate the parameter. The work is invariably easier with it than without it.**

38. Worked Examples.

(1) Find the area between $y = x^2 + x + 2$, $x = 4$, $x = 1$ and the axis of x.

Noticing first that the curve does not cross the axis we have

$$A = \int_1^4 (x^2 + x + 2)\, dx$$
$$= \left[\frac{x^3}{3} + \frac{x^2}{2} + 2x \right]_1^4$$
$$= \frac{63}{3} + \frac{15}{2} + 6$$
$$= 34\tfrac{1}{2} \text{ units.}$$

(2) Find the area included between the parabola $= 4ax$, the axis of x and the line $x = h$.

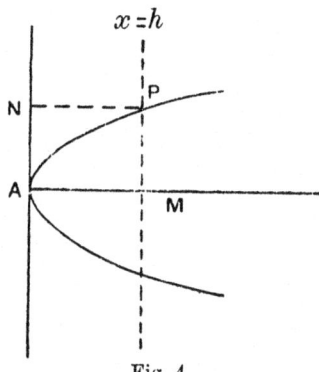

Fig. 4.

It is a little easier to calculate the area APN and to subtract it from the rectangle $AMPN$.

When $x = h$, $y = 2\sqrt{ah}$.

$$\text{Area } APN = \int_0^{2\sqrt{ah}} x\, dy$$
$$= \int_0^{2\sqrt{ah}} \frac{y^2}{4a}\, dy$$
$$= \frac{1}{4a} \left[\frac{y^3}{3} \right]_0^{2\sqrt{ah}}$$
$$= \frac{2}{3} h \sqrt{ah}.$$

But rectangle $AMPN = 2h\sqrt{ah}$.

$\therefore$ Portion $APM = \frac{4}{3}h\sqrt{ah}$.

It is, of course, not difficult to calculate APM directly, but the above method is instructive.

(3) To find the area of the ellipse

$$\frac{x^2}{a^2} + \frac{y^2}{b^2} = 1.$$

The curve consists of four equal quadrants.

The curve is expressible parametrically in the form

$$x = a\cos\theta,$$
$$y = b\sin\theta,$$

and the first quadrant is traced out by the variation of θ from 0 to $\dfrac{\pi}{2}$.

$\therefore$ Area of 1st quadrant

$$= \int_{\frac{\pi}{2}}^{0} y\frac{dx}{d\theta} \,.\, d\theta$$

$$= \int_{\frac{\pi}{2}}^{0} b\sin\theta\,.\,(-a\sin\theta)\,d\theta$$

$$= ab\int_{0}^{\frac{\pi}{2}} \sin^2\theta\,d\theta$$

$$= ab\,.\,\frac{1}{2}\,.\,\frac{\pi}{2} \qquad \{\text{Ch. V}\}$$

$$= \frac{\pi ab}{4}\,.$$

Hence the whole area of the ellipse is πab.

(4) To find the area between $y^2(a+x) = (a-x)^3$ and its asymptote.

The asymptote clearly is $x = -a$, for as $x \rightarrow -a$, $y \rightarrow \pm\infty$.

Also the curve cuts the axis where $x = a$, and $x \not> a$, for if $x > a$ or $< -a$, $y^2 < 0$.

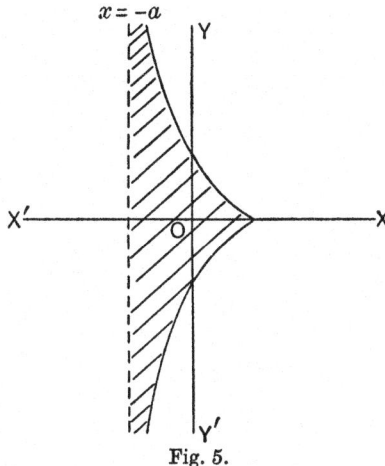

Fig. 5.

Hence the curve is as in Fig. 5.

Area is given by $2 \int_{-a}^{a} \dfrac{(a-x)^{\frac{3}{2}}}{(a+x)^{\frac{1}{2}}} dx$.

To evaluate put $x = a \sin \theta$.

The limits are now $\pm \dfrac{\pi}{2}$ and $dx = a \cos \theta \, d\theta$.

$$\therefore A = 2 \int_{-\frac{\pi}{2}}^{\frac{\pi}{2}} a^2 \frac{(1-\sin\theta)^{\frac{3}{2}}}{(1+\sin\theta)^{\frac{1}{2}}} . \cos\theta \, d\theta$$

$$= 2 \int_{-\frac{\pi}{2}}^{\frac{\pi}{2}} a^2 \frac{(1-\sin\theta)^{\frac{3}{2}}(1-\sin^2\theta)^{\frac{1}{2}}}{(1+\sin\theta)^{\frac{1}{2}}} d\theta$$

$$= 2a^2 \int_{-\frac{\pi}{2}}^{\frac{\pi}{2}} (1-\sin\theta)^2 \, d\theta$$

$$= 2a^2 \left[\theta + 2\cos\theta + \frac{\theta}{2} - \frac{\sin 2\theta}{4} \right]_{-\frac{\pi}{2}}^{\frac{\pi}{2}}$$

{expanding and integrating}

$$= 3\pi a^2 \text{ finally.}$$

39. Simpson's Approximate Rule for Area.

Consider the parabola $y = ax^2 + bx + c$.

The area included between this curve, the axis of x and two ordinates $x = h_1$, h_2, is clearly equal to

$$\frac{a}{3}\{h_2{}^3 - h_1{}^3\} + \frac{b}{2}\{h_2{}^2 - h_1{}^2\} + c\{h_2 - h_1\},$$

which may be written in the form

$$\{h_2 - h_1\}\left\{\frac{a}{3}\{h_2{}^2 + h_2 h_1 + h_1{}^2\} + \frac{b}{2}\{h_2 + h_1\} + c\right\}.$$

Now the ordinate at $x = h_1$ is clearly of length

$$ah_1{}^2 + bh_1 + c,$$

that at $x = h_2$ $ah_2{}^2 + bh_2 + c,$

and that at the mid-point $x = \dfrac{h_1 + h_2}{2}$

$$\frac{a}{4}\{h_1{}^2 + 2h_1 h_2 + h_2{}^2\} + \frac{b}{2}\{h_1 + h_2\} + c,$$

and we thus see that the area given above can be expressed as

$$\frac{h_2 - h_1}{6}\{1\text{st ordinate} + 4 \text{ times middle ordinate}$$

$$+ \text{2nd ordinate}\}\ldots(\mathbf{A}).$$

By considering every three consecutive points of a curve as if joined up by some parabola of the form given above (which we clearly have a right to do) we get Simpson's Rule for the Approximate Area under a curve. This is a result which is often useful in obtaining the area of a graph whose equation is not given.

The rule is as follows:

Divide the area by any <u>odd</u> number of equidistant ordinates. The area is then closely represented by the expression {one-third of interval between ordinates} × {1st ordinate + last ordinate + twice sum of all other odd ordinates + 4 times sum of all even ordinates}.

The student will easily prove this expression by repeating equation (A) the required number of times and adding.

EXAMPLES VI

(1) Find the area included between $y = 3x^2 + x + 3$, the axis of x and the ordinates $x = 2$, $x = 1$.

Also the area between the same curve, the axis of y and corresponding lines drawn perpendicular to it.

(2) Obtain the areas under the following curves between $x = 0$ and $x = h$:

$$(a) \quad y^3 = a^2 x;$$
$$(b) \quad a^2 y = x^3 + ax^2.$$

(3) Calculate the area of one arch of the curve $y = \sin x$.

(4) Find the total area included between the two parabolas $y^2 = 4ax$ and $x^2 = 4ay$.

(5) Use the Calculus to find the area of the circle $x^2 + y^2 = a^2$.

(6) Explain why the result of Ex. 5 cannot be regarded as a valid proof of the formula for the area of a circle.

(7) Find the area of the loop of the curve $y^2 = x^2 + x^3$.

(8) A cycloid is given by the equations

$$x = a\{\theta + \sin \theta\},$$
$$y = a\{1 - \cos \theta\}.$$

Make a rough figure of the curve, and calculate the area of one complete arch.

(9) Prove that the whole area of the curve $\left\{y - \dfrac{x^2}{a}\right\}^2 = a^2 - x^2$ is πa^2.

*(10) Find the ratio in which the curve $ay^2 = x^3$ divides the rectangle formed by drawing perpendiculars from any point on the two axes.

(11) Find the area above the axis of x bounded by the curve

$$y = (x^2 - 1)(4 - x^2).$$

(12) For the curve $y^2(a - x) = x^2(a + x)$ find

    (i) the area of the loop;

    (ii) the area between the curve and its asymptote.

(13) Find the whole area between $x^2 y^2 = a^2(y^2 - x^2)$ and its asymptotes.

CHAPTER VII

FURTHER APPLICATIONS TO GEOMETRY: SECTORIAL AREAS, VOLUMES OF REVOLUTION, FINDING THE LENGTH OF A CURVE

40. Areas in Polar Coordinates.

It is sometimes necessary to calculate areas belonging to curves whose equations are expressed in Polar Coordinates. The areas usually required are "sectorial areas," i.e. areas included between the curve and two given radii vectores.

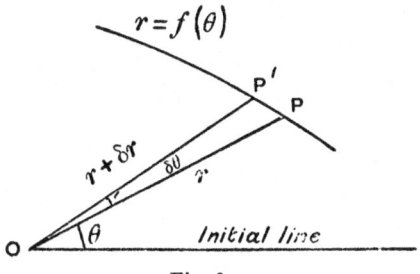

Fig. 6.

The formula for such areas can easily be found as follows: Consider two close points on the curve. In the limit when P and P' are on the point of coinciding we may regard the arc PP' as indistinguishable from its chord (cf. Fig. 5, which is a portion of a circle of large radius); if, then, P is (r, θ) and P' $(r + \delta r, \theta + \delta \theta)$,

the area $OPP' =$ area of $\triangle OPP' = \frac{1}{2} OP \cdot OP' \sin POP'$
$$= \frac{1}{2} r (r + \delta r) \sin \delta \theta.$$

But since a small angle and its sine are ultimately equal

this reduces to $\frac{1}{2}r(r+\delta r)\delta\theta$, and neglecting the term $\frac{1}{2}r\delta r\delta\theta$, which is of the second order, we have simply

$$\frac{1}{2}r^2\delta\theta.$$

Hence, since the area to be determined consists of a large number of such elements, it can by the fundamental theorem be found by evaluating $\dfrac{1}{2}\displaystyle\int_\alpha^\beta r^2 d\theta$, where $\theta=\beta$ and $\theta=\alpha$ are the two radii vectores which bound the area.

41. Example.

Consider the whole area of the curve $r=a(1+\cos\theta)$ [the Cardioid]. The curve consists of two equal portions

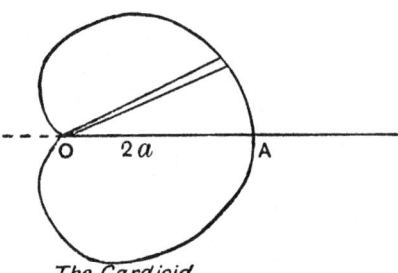

The Cardioid

Fig. 7.

as shown. The upper portion is traced by the revolution of the radius vector from 0 to π.

$$\therefore \text{ Area} = 2\int_0^\pi \tfrac{1}{2}r^2 d\theta$$

$$= \int_0^\pi a^2(1+\cos\theta)^2 d\theta$$

$$= a^2\int_0^\pi (1+2\cos\theta+\cos^2\theta)\,d\theta$$

$$= a^2\left[\theta+2\sin\theta+\frac{\theta}{2}+\frac{\sin 2\theta}{4}\right]_0^\pi$$

$$= \frac{3\pi a^2}{2}.$$

42. Volumes of Revolution.

The Calculus can also be employed to calculate the volumes of solids bounded by known surfaces. In general the process requires a knowledge of triple integration, but the simpler case of volumes of revolution can be dealt with by a single integration.

Consider the solid formed by the revolution about the axis of x of the portion of the curve $y = f(x)$ considered in § 33.

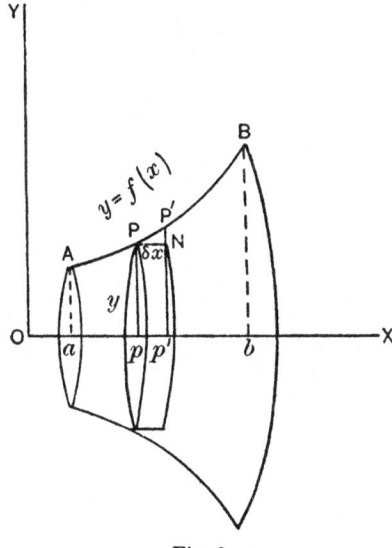

Fig. 8.

Dividing the area into strips as before we see that each strip, in revolving about OX, contributes to the volume a thin plate. In the limit we can regard $PP'p'p$ as a rectangle and the volume contributed by the revolution of this rectangle will be $\pi . Pp^2 . pp'$, or using the usual notation of the Calculus $\pi y^2 \delta x$.

Applying the fundamental theorem as before, we can find the sum of all these circular plates by evaluating $\int_a^b \pi y^2 dx$.

In the same way a volume of revolution about OY would be given by the formula $\int \pi x^2 dy$.

Also if x and y are given parametrically we can write the formula for a volume of revolution $\int \pi y^2 . \dfrac{dx}{dt} dt$ and evaluate in terms of the parameter t.

43. Worked Examples.

(1) To find the volume of a sphere.

Consider the revolution of one quadrant of the circle

$$x^2 + y^2 = a^2.$$

In the first quadrant x ranges from a to 0.

$$\therefore \text{Volume contributed} = \pi \int_0^a (a^2 - x^2)\, dx$$
$$= \pi \left\{ a^2 x - \frac{x^3}{3} \right\}_0^a$$
$$= \frac{2\pi a^3}{3}.$$

$\therefore$ Whole volume $= \frac{4}{3}\pi a^3$.

(2) To find the volume of a cone of height h and base radius r.

The cone can be considered as formed by the revolution of the line $y = \dfrac{r}{h} x$ about Ox.

$$\therefore \text{Volume} = \int_0^h \pi \frac{r^2}{h^2} x^2 \, dx$$

$$= \pi \frac{r^2}{h^2} \cdot \left(\frac{x^3}{3}\right)_0^h$$

$$= \frac{\pi r^2 h}{3}.$$

(3) To find the volume of revolution formed by the loop of the curve $y^2 = x^2 \dfrac{a - x}{a + x}$.

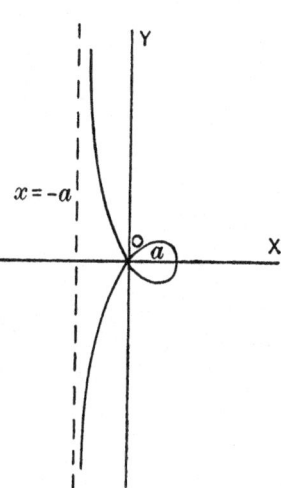

Fig. 9.

[The whole curve lies between $\pm a$, for otherwise y^2 is negative, also at the origin $y^2 = x^2$ and $y \to \pm \infty$ as $x \to -a$. Hence curve is as in figure.]

Volume traced by loop

$$= \pi \int_0^a y^2 \, dx$$

$$= \pi \int_0^a x^2 \cdot \frac{a - x}{a + x} \, dx.$$

Put $z = x + a$, then $dz = dx$ and new limits are $2a$ and a.

$\therefore$ Volume

$$= \pi \int_a^{2a} \frac{(z - a)^2 (2a - z)}{z} \, dz$$

$$= \pi \int_a^{2a} \left[\frac{2a^3}{z} - 5a^2 + 4az - z^2\right] dz$$

$$= \pi \left[2a^3 \log z - 5a^2 z + 2az^2 - \frac{z^3}{3}\right]_a^{2a} = 2\pi a^3 \left\{\log 2 - \tfrac{2}{3}\right\}.$$

EXAMPLES VII A

(1) Find the area of a loop of the curve $r = a \sin 4\theta$.

(2) Find the whole area of $r^2 = a^2 \cos^2 \theta + b^2 \sin^2 \theta$.

(3) Find the area of one loop of $r^2 = a^2 \cos 2\theta$. (The Lemniscate.)

(4) Find the area of the Limaçon $r = a + b \cos \theta$ when $a > b$.

(5) Modify the result of Ex. 4 to cover the cases (i) $b = a$, (ii) $a < b$.

(6) Find the area enclosed between the spiral $r = ae^\theta$, the initial line and the radius $\theta = \pi$.

(7) Find the volume of the "frustum" of a cone, if the height of the frustum is h and the radii of its ends r_1 and r_2.

(8) Find the volume cut off from a sphere of radius a by two parallel planes which meet the sphere in circles of radii r_1 and r_2.

(9) Find the volume of the "spheroid" traced out by the revolution of the ellipse $\dfrac{x^2}{a^2} + \dfrac{y^2}{b^2} = 1$.

(10) Assuming the earth's section to be an ellipse of eccentricity $\frac{1}{60}$, the minor axis being the line from the centre to the North Pole, calculate the percentage error in taking for its volume the volume of the sphere whose radius is that from the centre to the equator.

*(11) Find the volume formed by the revolution about Ox of the portion of the parabola $y^2 = 4ax$ bounded by $x = h$, and prove that this volume is $\frac{2}{3}$ of that of the enveloping cone formed by the tangents at the points where $x = h$.

(12) Find the volume formed by the revolution about Ox of the portion of the cycloid $x = a(\theta + \sin \theta)$, $y = a(1 - \cos \theta)$ given by values of θ ranging from 0 to π.

*(13) Find the volume of the segment of a sphere of radius r cut off by a plane at a distance $r - c$ from the centre.

*(14) A cylindrical hole is drilled symmetrically through a hemisphere at right angles to its plane surface. The radius of the hole being $\frac{3}{8}$ that of the sphere, find the volume of the portion removed.

*(15) A solid is generated by the revolution of $y^2 = Ax^2 + Bx + C$ about the axis of x. Prove that the volume cut off by two planes perpendicular to Ox at a distance h apart is $\frac{1}{6}h \{A_1 + 4A + A_2\}$, where A_1, A_2 are the end-areas and A the area of the section half-way along.

*(16) The segment of the parabola $y^2 = 4ax$ cut off by $x = h$ revolves about that ordinate. Prove that the volume so formed is $\frac{8}{15}$ of that of the cylinder formed by the revolution of the tangent at the vertex.

*(17) Find the position of a point on the curve $ay^2 = x^3$ such that, if perpendiculars are drawn from it to the axes, the volumes formed by the revolution of the two parts about their appropriate axes are equal.

44. Finding the Length of an Arc.

The Calculus may also be employed to find the length of an arc of any curve whose equation is given.

For if P and P' are two close points of the curve we may in the limit regard the chord PP' as indistinguishable from the arc PP'.

But chord

$$PP' = \sqrt{P'M^2 + PM^2},$$

i.e. $\delta s^2 = \delta x^2 + \delta y^2$.

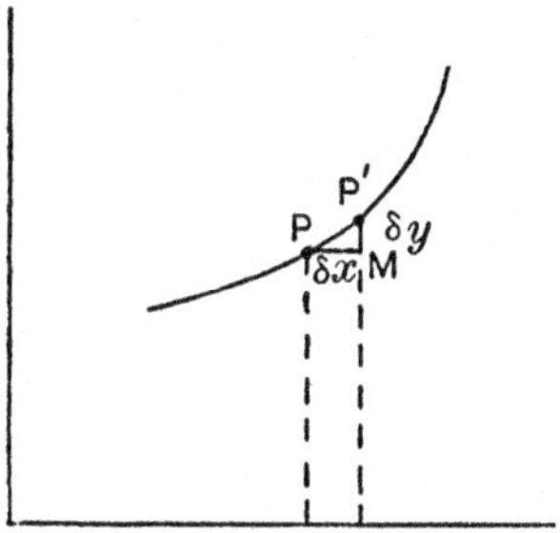

Fig. 10.

Various formulæ for s can at once be derived from this.

For considering x as independent variable we have, dividing by δx^2 and taking the limit,

$$\frac{ds}{dx} = \sqrt{1 + \left(\frac{dy}{dx}\right)^2},$$

$$\therefore s = \int \sqrt{1 + \left(\frac{dy}{dx}\right)^2}\, dx,$$

the integration being taken between limits corresponding to the ends of the arc to be measured.

Or, again, if y is considered as the primary variable we get

$$\frac{ds}{dy} = \sqrt{1 + \left(\frac{dx}{dy}\right)^2},$$

or

$$s = \int \sqrt{1 + \left(\frac{dx}{dy}\right)^2}\, dy.$$

Most useful of all is the "parametric" form

$$\left(\frac{ds}{dt}\right)^2 = \left(\frac{dx}{dt}\right)^2 + \left(\frac{dy}{dt}\right)^2,$$

whence

$$s = \int \sqrt{\left(\frac{dx}{dt}\right)^2 + \left(\frac{dy}{dt}\right)^2}\, dt.$$

It is assumed in all cases that the curve is continuous between the two ends.

45. Length. Formulæ in Polar Coordinates.

Or again, in Polar coordinates, if OP and OP' be two

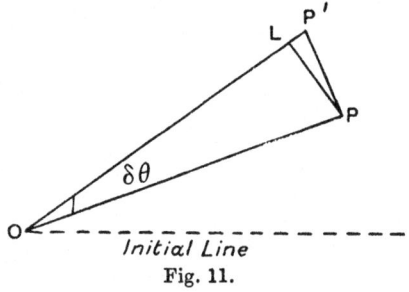

Fig. 11.

close radii vectores and PL perpendicular to OP', we have the arc PP' approximately equal to the chord PP'.

But

$$PP'^2 = PL^2 + LP'^2.$$

In the limit when $\delta\theta$ is a small angle we may consider
$$OL = OP,$$
$$\therefore\ LP' = r + \delta r - r = \delta r,$$
and also LP approx. = arc of a circle of radius OP
$$= r\,\delta\theta.$$
$$\therefore\ \delta s^2 = \delta r^2 + r^2 \delta\theta^2.$$

This in the same manner as in the preceding article gives us the results
$$s = \int \sqrt{1 + r^2\left(\frac{d\theta}{dr}\right)^2}\,dr,$$
and
$$s = \int \sqrt{r^2 + \left(\frac{dr}{d\theta}\right)^2}\,d\theta.$$

The process of finding s is often called " rectification," or " rectifying the arc."

46. Worked Examples.

(1) To find the length of the parabola $y^2 = 4ax$ included between the vertex and the point where $x = a$.

Choosing y as the primary variable the limits are $y = 2a,\ y = 0$.

$\therefore$ By formula
$$\begin{aligned}
s &= \int_0^{2a} \sqrt{1 + \left(\frac{dx}{dy}\right)^2}\,dy \\
&= \int_0^{2a} \sqrt{1 + \left(\frac{y}{2a}\right)^2}\,dy \qquad \left\{\text{for } \frac{dy}{dx} = \frac{2a}{y}\right\} \\
&= \frac{1}{2a} \int_0^{2a} \sqrt{y^2 + 4a^2}\,dy \\
&= \frac{1}{2a}\left[\frac{y\sqrt{y^2 + 4a^2}}{2} + \frac{4a^2}{2}\log\frac{y + \sqrt{y^2 + 4a^2}}{2a}\right]_0^{2a} \\
&= \frac{1}{4a}\left[2a\sqrt{8a^2} + \frac{4a^2}{2}\log(1 + \sqrt{2})\right] \\
&= a\{\sqrt{2} + \log(\sqrt{2} + 1)\}.
\end{aligned}$$

(2) Find the whole length of the cardioid

$$r = a\,(1 + \cos\,\theta).$$

We have noticed that the curve is symmetrical, the upper half being traced by the variation of θ from 0 to π.

Also
$$\frac{dr}{d\theta} = -\,a\sin\,\theta.$$

$$\begin{aligned}
\therefore \text{ Length} &= 2\int_0^\pi \sqrt{r^2 + \left(\frac{dr}{d\theta}\right)^2}\,d\theta \\
&= 2\int_0^\pi \sqrt{a^2\,(1 + \cos\,\theta)^2 + a^2\sin^2\theta}\,d\theta \\
&= 2a\int_0^\pi \sqrt{2 + 2\cos\,\theta}\,d\theta \\
&= 2a\int_0^\pi 2\cos\frac{\theta}{2}\,d\theta \\
&= \left[8a\sin\frac{\theta}{2}\right]_0^\pi \\
&= 8a.
\end{aligned}$$

The integrals given by these formulæ are often of great complexity: the rectification of so simple a curve as the ellipse is, indeed, not expressible without the invention of new functions. Only a few examples are given herewith; and for further information the student is referred to the standard text-books on the Calculus.

EXAMPLES VII B

(1) In the "catenary" $2y = c\left\{e^{\frac{x}{c}} + e^{-\frac{x}{c}}\right\}$, prove that the arc from $x = a$ to $x = 0$ is $\dfrac{c}{2}\left\{e^{\frac{a}{c}} - e^{-\frac{a}{c}}\right\}$.

(2) Find the length of any arc of the spiral $r = a\theta$.

(3) Find the length of a complete arch (from $\theta = 2\pi$ to $\theta = 0$) of the cycloid $x = a\,(\theta + \sin\,\theta)$, $y = a\,(1 - \cos\,\theta)$.

(4) Find the length of arc of the curve given oy

$$x = (a+b) \cos \theta - b \cos \frac{a+b}{b} \theta,$$

$$y = (a+b) \sin \theta - b \sin \frac{a+b}{b} \theta,$$

from the point $\theta = \dfrac{\pi b}{a}$ to the point $\theta = \theta_1$.

(5) Find the length from the origin to the point where $x = a$ in the curve $y = a \log \dfrac{a^2}{a^2 - x^2}$.

(6) Obtain a formula for the length of any arc of the curve $ay^2 = x^3$.

7) Find the whole length of the curve $x^{\frac{2}{3}} + y^{\frac{2}{3}} = a^{\frac{2}{3}}$.

CHAPTER VIII

APPLICATION TO PROBLEM OF FINDING CENTRES OF GRAVITY

47. Centre of Gravity. Principle.

We often require in Mechanics to determine the position of the centre of gravity of a given body, plane or solid. The coordinates of the C.G. can always be expressed by means of the Calculus, if the equation to the bounding curve or surface is known.

We shall confine our treatment to two cases, viz. Areas, and Surfaces of Revolution, since these are the only cases which do not in general involve double or triple integration.

The principle invoked is the Statical Principle of Moments, viz. that if a body be divided in any manner, the algebraical sum of the moments of the separate portions about any axis is equal to the moment about that axis of the whole body supposed collected at its C.G.

48. To determine the Centre of Gravity of an Area.

Consider first the problem of finding the C.G. of an area. Regard the area as a thin plane lamina of uniform surface density σ; and suppose that the area is that bounded by the curve $y = f(x)$, the axis of x and the two ordinates $x = a, x = b$. We have previously seen that any area can be obtained by adding or subtracting two or more such areas.

To determine the x coordinate of the C.G. we proceed as follows: Divide the area into strips parallel to the axis of y. The area of a typical strip will be $y \, \delta x$ and its mass $\sigma y \, \delta x$.

Hence the moment of such a strip about Oy will be $\sigma xy\,\delta x$, for every part of the strip is (to the first order) at a distance x from Oy.

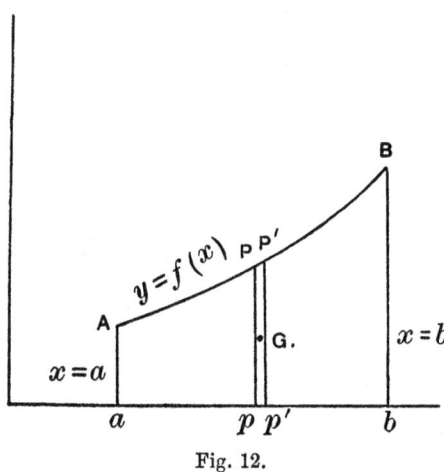

Fig. 12.

Applying the fundamental theorem the sum of all such moments will be $\int_a^b \sigma xy\,dx$.

But the whole area is $\int_a^b y\,dx$ and hence the whole mass $= \sigma \int_a^b y\,dx$.

Hence if $(\bar{x}, \bar{y})$ be the C.G. we have

$$\bar{x} \cdot \sigma \int_a^b y\,dx = \sigma \int_a^b xy\,dx,$$

i.e.

$$\bar{x} = \frac{\int_a^b xy\,dx}{\int_a^b y\,dx}.$$

Again, since any strip is approximately rectangular, its c.g. is at a height $\frac{y}{2}$ above Ox.

$\therefore$ Moment of any strip about $Ox = \sigma \dfrac{y^2}{2} \delta x$.

$\therefore$ Sum of all such moments $= \sigma \displaystyle\int_a^b \dfrac{y^2}{2}\, dx$.

To determine $\bar{y}$ we have the equation

$$\bar{y} \cdot \sigma \int_a^b y\,dx = \sigma \int_a^b \frac{y^2}{2}\,dx.$$

$$\therefore \bar{y} = \tfrac{1}{2} \frac{\displaystyle\int_a^b y^2\,dx}{\displaystyle\int_a^b y\,dx}.$$

The student will find it wise, at any rate in the earlier stages of his knowledge, to remember the method rather than the formula and to treat each example on its own merits *ab initio*.

49. We may, in just the same manner, find the c.g. of a volume of revolution. It is clear however in this case that the c.g. lies on the axis of revolution, and hence, in general, there is only one coordinate to be determined. To fix our ideas let us suppose that Ox is the axis of revolution, the solid is that considered in § 42 and the figure is Fig. 8.

The solid can, as in that article, be considered as formed of a large number of circular slabs each generated by the revolution of one of the elementary rectangles. Let ρ be the density.

The volume of any such slab $= \pi y^2 \delta x$.

$\therefore$ Its moment about $Oy = \pi \rho\, y^2 x\, \delta x$.

$\therefore$ Sum of all such moments $= \pi \rho \displaystyle\int_a^b y^2 x\, dx$.

But the whole mass $= \pi\rho \int_a^b y^2 dx$ and we thus get the equation

$$\bar{x} \cdot \pi\rho \int_a^b y^2 dx = \pi\rho \int_a^b xy^2 dx.$$

$$\therefore \ \bar{x} = \frac{\displaystyle\int_a^b xy^2 dx}{\displaystyle\int_a^b y^2 dx}.$$

It is seen that the method can easily be extended to cover cases of variable density, as long as the density involves x only. And again it is the method that is worth remembering, not the formula.

50. This is by far the easiest way of finding the C.G. of the simpler bodies as the following examples will show.

(i) Find the C.G. of the area between $y^2 = 4ax$, $y = 0$. and $x = a$.

Dividing into strips parallel to Oy, the total moment of the strips is seen to be

$$\sigma \int_0^a xy \, dx$$

$$= \sigma \int_0^a x \cdot 2\sqrt{ax}\, dx$$

$$= 2\sigma\sqrt{a} \int_0^a x^{\frac{3}{2}} \, dx$$

$$= \tfrac{4}{5}\sigma a^3.$$

But the whole mass $= \sigma \int_0^a y\, dx$

$$= 2\sigma\sqrt{a} \int_0^a x^{\frac{1}{2}} dx$$

$$= \tfrac{4}{3}\sigma a^2.$$

$$\therefore \ \tfrac{4}{3}\sigma a^2 \cdot \bar{x} = \tfrac{4}{5}\sigma a^3,$$

$$\therefore \ \bar{x} = \tfrac{3}{5}a.$$

6

Again, since the ordinate of the C.G. of each strip is $\frac{y}{2}$ the moment of all the strips about Ox is

$$\sigma \int_0^a \frac{y^2}{2} dx$$
$$= \tfrac{1}{2}\sigma \int_0^a 4axdx$$
$$= \sigma a^3.$$
$$\therefore \tfrac{4}{3}\sigma a^2 . \bar{y} = \sigma a^3,$$
$$\therefore \bar{y} = \frac{3a}{4}.$$

The C.G. is thus $\left\{ \dfrac{3a}{5}, \ \dfrac{3a}{4} \right\}$.

(ii) Find the C.G. of a solid cone.

The cone can, as we have seen, be traced out by the revolution of the line $y = \dfrac{r}{h} x$.

Dividing the cone into circular slabs perpendicular to Ox, the moment of any slab about $Oy = \pi \rho xy^2 \delta x$.

$\therefore$ Total moment of all such slabs

$$= \int_0^h \pi \rho xy^2 dx$$
$$= \pi \rho \frac{r^2}{h^2} \int_0^h x^3 dx$$
$$= \pi \rho \frac{r^2}{h^2} \cdot \frac{h^4}{4}$$
$$= \tfrac{1}{4}\pi \rho r^2 h^2.$$

But we have proved that the volume $= \tfrac{1}{3}\pi r^2 h$.

$$\therefore \tfrac{1}{3}\pi \rho r^2 h . \bar{x} = \tfrac{1}{4}\pi \rho r^2 h^2,$$
$$\therefore \bar{x} = \frac{3h}{4},$$

i.e. the C.G. is one quarter of the way up the axis of the cone.

(iii) To find the c.g. of a segment of a paraboloid of revolution formed by the revolution of $y^2 = 4ax$ about Ox, the density at any point varying as x^2, i.e. $\rho = kx^2$.

Proceeding as before

the mass   of a slab is $\pi k x^2 y^2 \delta x$,

and the moment of a slab is $\pi k x^3 y^2 \delta x$.

Hence
$$\bar{x} . \int_0^h \pi k x^2 y^2 \, dx = \int_0^h \pi k x^3 y^2 \, dx,$$

i.e.
$$\bar{x} . \int_0^h 4ax^3 \, dx = \int_0^h 4ax^4 \, dx,$$

i.e.
$$\bar{x} . h^4 = \frac{4h^5}{5}.$$

$$\therefore \bar{x} = \frac{4h}{5}.$$

(iv) To find the surface and c.g. of a hollow hemisphere.

Consider the revolution of an element of arc of a quadrant of the circle $x^2 + y^2 = a^2$.

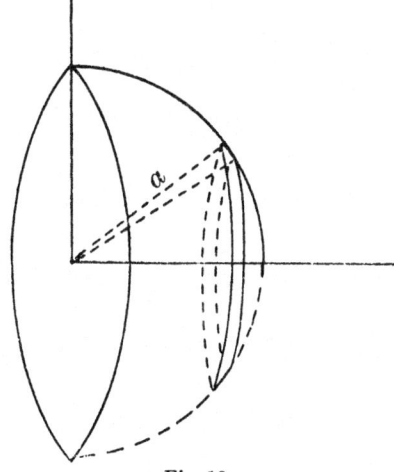

Fig. 13.

The element of arc traces out a belt, and the sum of all such belts is the surface of the hemisphere.

Clearly the radius of the circle traced out by the belt at a point (x, y) is y.

And hence the area of such a belt is $2\pi y \,.\, \delta s$. Hence the whole area is $\int 2\pi y \, ds$ taken over limits which include the whole quadrant.

For the circle $x = a \cos \theta$, $y = a \sin \theta$.

Hence $$\frac{ds}{d\theta} = \sqrt{\left(\frac{dx}{d\theta}\right)^2 + \left(\frac{dy}{d\theta}\right)^2} = a,$$

i.e. $$ds = a \, d\theta,$$

and the whole quadrant is covered if θ varies from 0 to $\dfrac{\pi}{2}$.

$$\therefore \text{ Integral becomes} \int_0^{\frac{\pi}{2}} 2\pi a \sin \theta \,.\, a \, d\theta$$
$$= 2\pi a^2 \int_0^{\frac{\pi}{2}} \sin \theta \, d\theta$$
$$= \mathbf{2\pi a^2}.$$

Again, to find the C.G.; clearly the C.G. of any belt is at its centre.

$\therefore$ Moment of all belts

$$= \sigma \int 2\pi yx \, ds$$
$$= 2\pi\sigma \int_0^{\frac{\pi}{2}} a^3 \sin \theta \cos \theta \, d\theta$$
$$= \pi\sigma a^3 \int_0^{\frac{\pi}{2}} \sin 2\theta \, d\theta$$
$$= \pi\sigma a^3 \left[-\tfrac{1}{2} \cos 2\theta \right]_0^{\frac{\pi}{2}}$$
$$= \pi\sigma a^3.$$

$\therefore$ We have $2\pi\sigma a^2 . \bar{x} = \pi\sigma a^3$.

Hence $\qquad \bar{\mathbf{x}} = \dfrac{\mathbf{a}}{\mathbf{2}}$ and clearly $\bar{\bar{y}} = 0$.

51. Pappus' Theorems.

The following are two theorems often attributed to Guldinus but also found in the works of the Greek geometer Pappus.

(i) *If any closed area revolves about an axis external to the area, the volume formed is equal to the area multiplied by the length of the path traced out by the* C.G. *of the area.*

(ii) *The surface area so formed is equal to the perimeter of the enclosing curve multiplied by the path of the* C.G.

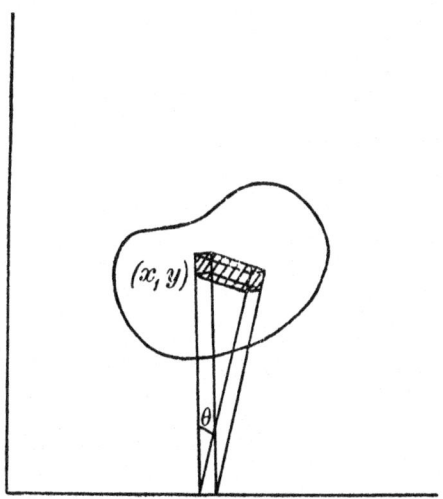

Fig. 14.

A proof of the first of these theorems will suffice.

Consider an element of the area, δA, and suppose that it rotates through an angle $\delta\theta$ about Ox.

The arc traced out by any point within the element will be to the first order $y\,\delta\theta$ and hence the volume contributed by the element is $y\,\delta A\,\delta\theta$.

Or in fact for a rotation through a finite angle α the element will be $y\alpha\,\delta A$, and thus the whole volume will be given by $\int y\alpha\,dA$, the integration being taken over such limits as will cover the whole area.

But, by our previous work, the ordinate of the c.g. can be found from the equation

$$\bar{y} \times \text{Total Mass} = \text{Total Moment},$$

i.e. $$\bar{y} \times \int dA = \int y\,dA.$$

$$\therefore \bar{y} \cdot A = \int y\,dA.$$

$\therefore$ Volume $= \int y\alpha\,dA = \alpha\int y\,dA = \bar{y}\alpha \cdot A$, and $\bar{y}\alpha$ is the path traced by the c.g.

Therefore the theorem is as stated.

In particular for a whole revolution $V = 2\pi\bar{y} \cdot A$.

It is seen that a similar proof gives the second theorem.

52. Volumes of Revolution. Polar Coordinates.

Pappus' Theorem can be employed to give the volume of revolution formed by an area given in polar coordinates. For we have previously seen that an element of area is $\frac{1}{2}r^2\,\delta\theta$, and, regarding this as a triangle, its c.g. is $\frac{2}{3}$ of the way along the median, i.e. at a point whose coordinates are approximately $\left\{\dfrac{2r\cos\theta}{3},\ \dfrac{2r\sin\theta}{3}\right\}$.

$\therefore$ Path of c.g. $= \frac{4}{3}\pi r\sin\theta$.

∴ Volume contributed by an element $= \frac{2}{3}\pi r^3 \sin\theta\, \delta\theta$.

∴ Whole volume is given by $\int \frac{2}{3}\pi r^3 \sin\theta\, d\theta$ between suitable limits.

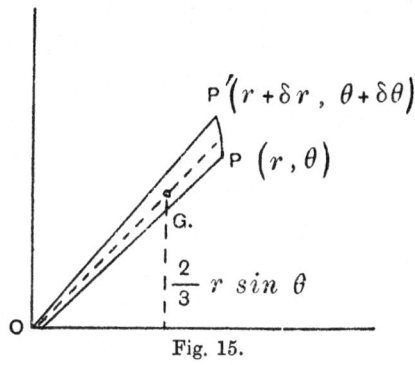

Fig. 15.

Ex. Find the volume traced out by revolution of
$$r = a + b\cos\theta \quad (a > b)$$
about the initial line.

As above
$$V = \tfrac{2}{3}\pi \int_0^\pi r^3 \sin\theta\, d\theta$$
$$= \tfrac{2}{3}\pi \int_0^\pi (a + b\cos\theta)^3 \sin\theta\, d\theta.$$

Let
$$u = a + b\cos\theta,$$
$$du = -b\sin\theta\, d\theta.$$

And limits for u are $a - b,\ a + b$.

$$\therefore V = \frac{2\pi}{3} \int_{a+b}^{a-b} -u^3 \cdot \frac{du}{b}$$
$$= \frac{2\pi}{3b} \left\{ \frac{(a+b)^4}{4} - \frac{(a-b)^4}{4} \right\}$$
$$= \frac{2\pi}{12b} \cdot [2a^2 + 2b^2]\,[4ab]$$
$$= \frac{4\pi}{3} a\,(a^2 + b^2).$$

If $a < b$, the limits for θ are 0 and $\cos^{-1}\left(-\dfrac{a}{b}\right)$.

EXAMPLES VIII

Find the c. g. of:

(1) a semi-circle;

(2) the circumference of a semi-circle;

(3) a solid hemisphere;

(4) a quadrant of the ellipse $\dfrac{x^2}{a^2} + \dfrac{y^2}{b^2} = 1$.

(5) Show that the area bounded by the curve $r = f(\theta)$ and the radii vectores $\theta = a$, $\theta = \beta$ has its c. g. at the point

$$x = \frac{2}{3} \frac{\displaystyle\int_a^\beta r^3 \cos\theta\, d\theta}{\displaystyle\int_a^\beta r^2\, d\theta}, \quad y = \frac{2}{3} \frac{\displaystyle\int_a^\beta r^3 \sin\theta\, d\theta}{\displaystyle\int_a^\beta r^2\, d\theta}.$$

(6) Prove that the c. g. of the area enclosed by the cardioid $r = a(1 + \cos\theta)$ is at $\left(\dfrac{5a}{6},\, 0\right)$.

(7) Find the c. g. of the area between $y^2 = 4ax$ and $y = mx$.

(8) Find the position of the c. g. of the portion of a sphere intercepted between two parallel planes.

Find also an expression for the surface-area of this zone.

(9) Find the c. g. of a frustum of a solid cone, h being the height of the frustum and r, R the radii of its ends.

(10) Find the surface and c. g. of a hollow cone.

(11) Find the c. g. of a straight rod of length l in which the density of an element is proportional to the distance of the element from one end of the rod.

(12) The part of the curve $y^{m+n} = a^m x^n$ between $x = 0$ and $x = h$ revolves about Ox. Find the c. g. of the solid so formed.

(13) Give expressions for the volume and total surface of the "anchor-ring" formed by the revolution of a circle of radius a about an axis in its plane distant d from its centre.

(14) A square revolves about a line through one corner parallel to a diagonal. Find the volume and total surface so formed.

(15) Find the volume traced out by the revolution about the initial line of the loop of the curve $r^3 = a^3 \theta \cos \theta$.

(16) The area between the cardioid $r = 2a (1 + \cos \theta)$ and the parabola $r (1 + \cos \theta) = 2a$ revolves about the initial line. Prove that the volume formed is $18\pi a^3$.

(17) A cotton reel is formed by the revolution of the given figure

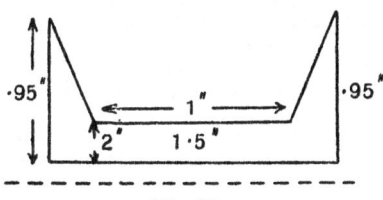

Fig. 16.

about an axis $\frac{1}{10}''$ from its straight edge. Find the volume and total surface of the reel.

*(18) In Ex. 15, VII A, prove that the C.G. is at a distance

$$\frac{h}{2} \frac{A_2 - A_1}{A_2 + 4A + A_1}$$

from one end.

(19) By considering the revolution of a semi-circular area about a diameter find the position of the C.G. of a semi-circle by Pappus' Theorem.

(20) By considering the surface formed by the revolution of one quarter of the circumference of a circle, deduce the C.G. of one quarter of the circumference of a circle.

*(21) An arc of a circle of radius c subtending an angle $2a$ at the centre is rotated about its chord. Prove that the area of the surface of revolution so formed is $4\pi c^2 (\sin a - a \cos a)$.

(22) Deduce from Ex. 21 the position of the C.G. of any circular arc.

CHAPTER IX

FURTHER APPLICATIONS TO MECHANICS. RIGID DYNAMICS

53. Moments of Inertia.

If a particle of mass m lies at a distance r from a fixed axis, the product mr^2 is known as the "moment of inertia" of the particle about that axis. This quantity, as we shall see, figures largely in questions of rotation of a heavy body, and the calculation of moments of inertia is thus one of the most important practical applications of the Calculus. In general we are dealing with solid bodies and not with particles, and we thus have to consider our rigid bodies as composed of a large number of strips, or other convenient divisions, and obtain the sum of the resulting series of quantities by the Calculus. Only in some of the simpler cases may this be effected by a single integration.

There are two general theorems relating to moments of inertia which are often of use in the evaluation and which we proceed to give.

54. Two Important Results on Moments of Inertia.

A. If I_x and I_y are the moments of inertia about two perpendicular axes OX and OY, of a plane lamina lying in the plane XOY, and I_z the moment of inertia of the same lamina about OZ which is perpendicular to the plane XOY, then

$$I_z = I_x + I_y.$$

For consider an element of the body of mass m, lying at a point (x, y) on the plane.

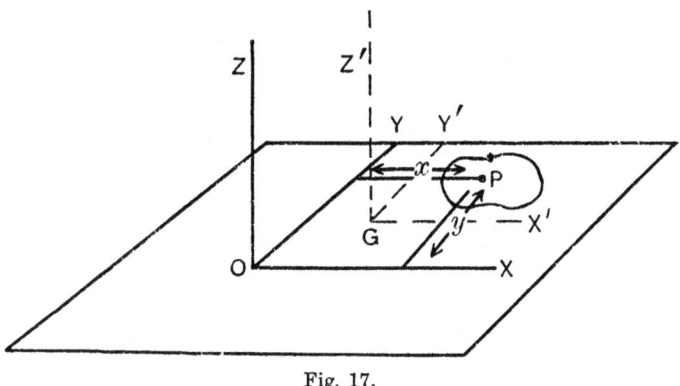

Fig. 17.

Then moment of inertia of the element about OX
$$= mx^2,$$
and that about $OY \qquad = my^2,$
i.e. $I_x = \Sigma mx^2$ and $I_y = \Sigma my^2$, summing the elements.
$$\therefore \ I_x + I_y = \Sigma m (x^2 + y^2) = \Sigma m \cdot OP^2$$
$$= I_z,$$
$$\therefore \ I_z = I_x + I_y.$$

B. If I_A is the moment of inertia of any body of mass M about any axis and I_G the M.I. about a parallel axis through the C.G., these two axes being a distance h apart, then
$$I_A = I_G + Mh^2.$$

First suppose the body is a lamina as in the previous theorem and OZ is the axis under consideration, and let the C.G. be at $(\bar{x}, \bar{y})$.

If now parallel axes GX', GY', GZ' are taken through

G and (x, y) are the old, (x', y') the new coordinates of P, we have
$$x = x' + \bar{x},$$
and
$$y = y' + \bar{y}.$$

Now the moment of inertia about OZ

$$= \Sigma m (x^2 + y^2)$$
$$= \Sigma m (x'^2 + 2x'\bar{x} + \bar{x}^2 + y'^2 + 2y'\bar{y} + \bar{y}^2)$$
$$= \Sigma m (x'^2 + y'^2) + \Sigma m (\bar{x}^2 + \bar{y}^2) + 2\Sigma m x'\bar{x} + 2\Sigma m y'y$$
$$= \Sigma m (x'^2 + y'^2) + (\bar{x}^2 + \bar{y}^2) \Sigma m + 2\bar{x} \Sigma m x' + 2\bar{y} \Sigma m y',$$

bringing outside certain constant terms.

Now $\Sigma m (x'^2 + y'^2)$ is clearly the moment of inertia about the axis GZ'; also $\bar{x}^2 + \bar{y}^2 = OG^2 = h^2$.

The two terms $\Sigma m x'$, $\Sigma m y'$ must both vanish for they clearly represent the moments of the lamina about the two axes GX', GY' which contain the c.g.

Hence we get

$$I_{OZ} = I_{GZ} + Mh^2.$$

Considering a solid body as made up of a number of parallel laminæ we see that the theorem is also true for a solid body.

55. Reason for existence of "Moments of Inertia."

It is of interest at this stage to consider how the Moment of Inertia comes to play the important part it does in Mechanics. Consider a rigid body of any sort rotating about a fixed axis. Any particle of the body at a distance r from the centre will be describing a circle of radius r.

The tangential velocity of this particle will thus be $r\dfrac{d\theta}{dt}$ (θ is the angle described) and hence the tangential acceleration will be the derivative of this, viz. $r\dfrac{d^2\theta}{dt^2}$.

The "effective" force along the tangent will thus be $mr\dfrac{d^2\theta}{dt^2}$ and the moment of this force $mr^2\dfrac{d^2\theta}{dt^2}$.

Now by d'Alembert's Principle the sum of all such moments must be equal to the moment of the external forces which are giving rise to the motion. Let this moment be denoted by L and we now have the equation

$$\Sigma mr^2\frac{d^2\theta}{dt^2} = L.$$

But for any rigid body $\dfrac{d^2\theta}{dt^2}$ is constant for all particles composing the body and may be taken outside the summation sign, and Σmr^2 is the quantity we call I (the moment of inertia); we thus arrive at the fundamental equation for the rotation of a fixed body

$$\mathbf{I\,\frac{d^2\theta}{dt^2} = L.}$$

56. Analogies between Linear and Rotational Dynamics.

The student will notice the close analogy between this equation and that of ordinary Dynamics, $P = mf$.

In fact we may say in general that, in Rotational Dynamics, I takes the place of m in the same way that angular velocity and acceleration take the place of linear velocity and acceleration. For instance it is easily verified that $\frac{1}{2}I\left(\dfrac{d\theta}{dt}\right)^2$ is the energy possessed by a body in virtue of its rotation, and $I\dfrac{d\theta}{dt}$ its "angular" momentum.

Corresponding to the two equations of ordinary Dynamics:

(a) Change of K.E. = Force × Space described,

(b) Change of Momentum = Force × Time,

we have two similar equations in Rigid Dynamics:

(a) Change in value of $\frac{1}{2} I \left(\frac{d\theta}{dt}\right)^2$ = Moment of couple

producing change × Angle (in radians) turned through,

(b) Change in value of $I \frac{d\theta}{dt}$ = Moment of couple × Time.

By means of these equations most of the elementary questions in Rigid Dynamics can be solved. An example is given later.

57. Worked Examples.

A few worked examples are given here. For fuller information the student is referred to standard works on Rigid Dynamics.

(1) Find the M.I. of a uniform rod of length $2a$ about its centre.

Let λ be the mass of the rod per unit length. Consider an element of the rod at a distance x from O.

The M.I. of this element about its centre is $\lambda x^2 \delta x$.

Hence by the principles of the Calculus we have to find

$$\int_{-a}^{a} \lambda x^2 dx,$$

$$\therefore \ I = \left[\frac{\lambda x^3}{3}\right]_{-a}^{a} = \frac{2\lambda a^3}{3}.$$

But $2\lambda a = M$ (the mass of the rod),

$$\therefore \ I = \frac{Ma^2}{3}.$$

The quantity k which is such that the moment of inertia of the body $= Mk^2$ is often called the "radius of gyration"; e.g. in the last example the radius of gyration of the rod about its centre is $\dfrac{a}{\sqrt{3}}$ *.

(2) We can see similarly (using Theorem B) that the M.I. about one end is $\frac{4}{3}Ma^2$, and that the M.I. about an axis perpendicular to the rod and distant b from its mid-point is

$$M\left(\frac{a^2}{3} + b^2\right).$$

(3) The M.I. of a circular area about any diameter.

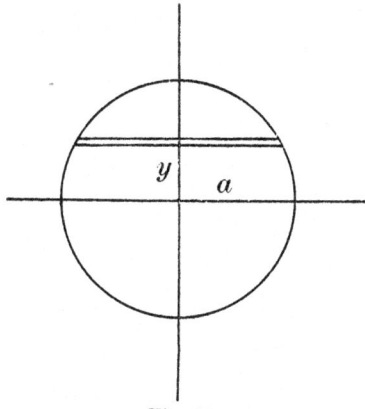

Fig. 18.

Taking the diameter as axis of x the length of a strip at a distance y from Ox is $2\sqrt{a^2 - y^2}$, and since every portion of this strip is at a distance y from Ox, the M.I. of the strip will evidently be $2y^2\sqrt{a^2 - y^2} \cdot \lambda \cdot dy$. ($\lambda =$ density.)

$$\therefore \text{ Whole M.I.} = 2\int_{-a}^{a} y^2 \sqrt{a^2 - y^2}\, dy.$$

* Physically it is the distance at which a particle of equal mass must be placed to rotate in the same manner as the given body.

Put

$$y = a \sin \theta,$$
$$dy = a \cos \theta \, d\theta,$$

and the new limits are $\pm \dfrac{\pi}{2}$.

$$\therefore \text{ M.I.} = 2 \int_{-\frac{\pi}{2}}^{\frac{\pi}{2}} a^4 \sin^2 \theta \cos^2 \theta \, d\theta$$

$$= \frac{\lambda a^4}{2} \int_{-\frac{\pi}{2}}^{\frac{\pi}{2}} \sin^2 2\theta \, d\theta$$

$$= \frac{\lambda a^4}{4} \int_{-\frac{\pi}{2}}^{\frac{\pi}{2}} (1 - \cos 4\theta) \, d\theta$$

$$= \frac{\lambda a^4}{4} \left[\theta - \frac{\sin 4\theta}{4} \right]_{-\frac{\pi}{2}}^{\frac{\pi}{2}}$$

$$= \frac{\pi \lambda a^4}{4}.$$

But

$$M = \pi \lambda a^2.$$
$$\therefore \text{ M.I.} = \frac{M a^2}{4}.$$

It follows by symmetry that the Moment of Inertia about Oy also equals $\dfrac{M a^2}{4}$. And since, by Theorem A, $I_z = I_x + I_y$, we find the Moment of Inertia of a circle about an axis through its centre perpendicular to its plane to be $\dfrac{M a^2}{2}$.

(4) The student should carefully notice the following worked example of a Dynamical character:

A uniform wheel of mass 100 lbs. and radius of gyration about its centre 1 foot is acted upon by a couple whose moment = 10 ft. lb. units for one minute. Find the angular velocity produced. Find also the constant couple which would stop the wheel in half a minute when it is revolving at 15 revs. per second and how many revolutions would be made before coming to rest.

$$\text{M.I.} = Mk^2 = 100 \text{ and } L = 10g \text{ units.}$$

$$\therefore \text{ the equation is } 100\frac{d^2\theta}{dt^2} = 10g.$$

Integrating both sides $100\frac{d\theta}{dt} = 10gt + C$, an equation giving the angular velocity at time t.

Clearly this velocity must be zero when $t = 0$.

$$\therefore C = 0 \text{ and } \frac{d\theta}{dt} = \frac{gt}{10}.$$

$\therefore$ when $t = 60$ secs. $\frac{d\theta}{dt} = 6g$, i.e. an angular velocity of $6g$ radians per sec. is produced.

Again, in the second part of the question, let $-L$ be the required retarding couple.

$$\therefore 100\frac{d^2\theta}{dt^2} = -L,$$

and integrating $100\frac{d\theta}{dt} = -Lt + C.$

But at zero time

$$\frac{d\theta}{dt} = 30\pi \quad (15 \text{ revs. per sec.}),$$

$$\therefore C = 3000\pi.$$

$$\therefore 100\frac{d\theta}{dt} = -Lt + 3000\pi.$$

If $\frac{d\theta}{dt}$ becomes zero at end of 30 secs. we have

$$30L = 3000\pi,$$

i.e. $\qquad\qquad L = 100\pi \text{ units,}$

i.e. required retarding couple is one whose moment is $\dfrac{100\pi}{g}$ ft. lbs.

For the last part of the question we have

$$100\,\frac{d^2\theta}{dt^2} = -100\pi,$$

i.e.

$$\frac{d^2\theta}{dt^2} = -\pi.$$

Multiply both sides by $\dfrac{d\theta}{dt}$ and integrate with respect to t,

i.e.

$$\frac{d\theta}{dt}\cdot\frac{d^2\theta}{dt^2} = -\pi\,\frac{d\theta}{dt},$$

and

$$\frac{1}{2}\left(\frac{d\theta}{dt}\right)^2 = -\pi\theta + C.$$

But when $\dfrac{d\theta}{dt} = 30\pi$, $\theta = 0$, for we are measuring the angle from the initial position given.

$$\therefore \frac{900\pi^2}{2} = C.$$

$$\therefore \frac{1}{2}\left(\frac{d\theta}{dt}\right)^2 = -\pi\theta + \frac{900\pi^2}{2}.$$

$\therefore$ when the wheel comes to rest, $\dfrac{d\theta}{dt} = 0$ and $\theta = 450\pi$,

i.e. the wheel has revolved 225 times.

The student should observe particularly the last method of integrating the given equation, which is often of use.

EXAMPLES IX A

Answers should be expressed in the form Mk^2.

(1) Find the Moment of Inertia of a rectangle whose sides are $2a$, $2b$, about axes through its centre parallel to its sides.

(2) Hence deduce the Moment of Inertia of a rectangular solid whose sides are $2a$, $2b$, $2c$, about principal axes through its centre.

Find the Moment of Inertia of each of the following:

(3) The circumference of a circle about a diameter.

(4) A thin hollow sphere about any diameter.

(5) An ellipse about its major and minor axes.

(6) A right circular cylinder about its axis.

(7) A circular disc about any tangent.

(8) A thin circular ring about any tangent.

(9) An isosceles triangle about the principal altitude.

(10) A hollow spherical shell of external radius b and internal radius a about a diameter.

(11) A flywheel of mass 500 lbs. has a radius of gyration of $2\frac{1}{2}$ feet. What is its energy when rotating at 200 revs. per minute? If its motion is opposed by a couple whose moment is 12 ft. lbs., find how long will elapse before the wheel comes to rest and how many revolutions it will have made.

(12) Find in the previous question the time in which the velocity will be reduced to half its initial value.

(13) A uniform trap-door of mass M and dimensions 2 feet by 3 feet turns about its smaller edge from an initially horizontal position. Write down its kinetic and potential energies when it has fallen through an angle θ, and deduce the velocity with which it reaches its lowest position.

58. Work done by a varying force.

The Calculus can be employed to find the work done by a varying force, when the law of variation of the force is known, and can be expressed in terms of some parameter x; for let δs be the displacement in the direction of the line of action of the force, corresponding to a small increment δx in the value of the parameter. The work done during this displacement is $P\delta s$ and hence the whole work can be calculated by finding $\int Pds$, the limits being such as to include the whole interval which is being considered.

As an example let us consider the problem of finding the work done in stretching an elastic string from its natural length a to a new length b.

When the length of the string is x $(x > a)$ by Hooke's Law the tension will be $\lambda \dfrac{x-a}{a}$, where λ is some constant depending on the composition of the string.

In stretching to a length $x + \delta x$, the tension may be considered as remaining constant and the work done is $\lambda \dfrac{x-a}{a} \delta x$.

$$\therefore \text{ Total work done} = \int_a^b \lambda \frac{x-a}{a}\, dx$$
$$= \frac{\lambda}{a}\left[\frac{(x-a)^2}{2}\right]_a^b$$
$$= \frac{\lambda}{2a}(b-a)^2.$$

But the final tension at length $b = \lambda \dfrac{b-a}{a}$.

Hence the work done $= \frac{1}{2}$ Extension $\times$ Final Tension.

59. Practical Application.

In practical work this calculation of work done is often performed graphically. We can see that if we have a graph in which Force is plotted against the distance the point of application has moved, the area under this curve will represent the work done between any two positions of the machine, for work done $= \int_{x=a}^{x=b} P\, dx =$ area of the $(P,\ x)$ curve between $x = b$, $x = a$.

For example, in the question considered in the preceding article $T = \lambda \dfrac{x-a}{a}$.

The graph would thus be a straight line inclined at an angle $\tan^{-1}\lambda$ and passing through $(a, 0)$. The work done is thus represented by the area of the $\triangle ATB$ and is clearly $\frac{1}{2}TB \cdot AB$, or $\frac{1}{2}$ Final tension × Extension. Devices are often introduced by which the (P, x) curve can be drawn

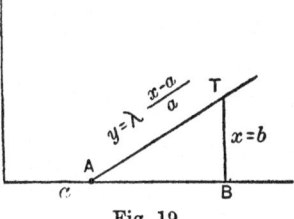

Fig. 19.

automatically by the machine performing the work and the area of these diagrams gives the work done.

Such devices are generally used with machines which perform a "cycle"; doing useful work in one part of the cycle, and requiring work done on them during the other part to restore the machine to its initial position. In a steam engine for instance work is done by the piston during part of the revolution but against the piston in restoring it to its original position.

The work diagram (known as an Indicator diagram) would be roughly of the shape shown in the sketch. The area $PAQNM$ represents the positive work done in the stroke, the area $PA'QNM$ the (negative) work taken up in restoring the piston to the

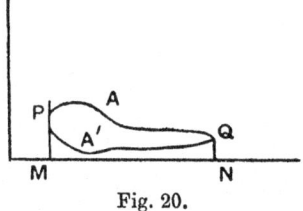

Fig. 20.

zero position. Hence the closed area $PAQA'$ represents the total effective work done per stroke.

60. Total Resultant Thrust and Centre of Pressure.

The Calculus is also used in hydrostatical problems. The pressure of a liquid is a variable force—proportional to the distance below the free surface of the body, or part of a body, on which the pressure acts. It is often desirable to calculate the Total Force (Total Resultant Thrust) on an unequally immersed body—and also to find the point (Centre of Pressure) at which this Total Thrust may be assumed to act.

Simple cases are obviously able to be evaluated by one integration, though more complicated cases often require double integration. The principle is again the Principle of Moments, moments being taken about the free surface of the liquid. A simple example is given herewith.

Ex. A square of side a is immersed vertically, with one edge in the free surface. Find the Total Resultant Thrust and the Centre of Pressure.

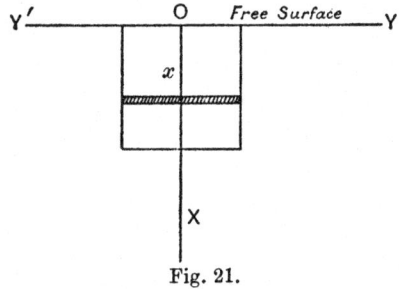

Fig. 21.

Taking the axis of y in the free surface, consider a strip parallel to the free surface at a depth x below it.

The area of the strip is $a\,\delta x$ and all along it the pressure will be $\rho g x$ units of force per unit of area.

$$\therefore \text{Total Thrust on strip} = \rho g\,ax\,\delta x.$$
$$\therefore \text{Total Thrust on square} = \int_0^a \rho g\,ax\,dx$$
$$= \frac{\rho g a^3}{2}.$$

Now the moment of the thrust on the strip about Oy

$$= \rho g \, a x^2 \delta x.$$

$$\therefore \text{Total moment} = \int_0^a \rho g a x^2 dx$$

$$= \frac{\rho g a^4}{3}.$$

$\therefore$ If Resultant thrust acts at a depth $\bar{x}$ we have

Moment of total thrust $=$ Total moment of thrusts on strips,

i.e.
$$\frac{\rho g a^3}{2} \bar{x} = \frac{\rho g a^4}{3},$$

$$\therefore \bar{x} = \frac{2a}{3}.$$

The centre of pressure is at $\frac{2}{3}$ of the depth.

EXAMPLES IX B

(1) A curve is drawn showing the velocity of a particle against the time. Explain how to read from the graph the distance described in a given interval.

Apply your result to the formula $v = u + ft$ and deduce graphically $s = ut + \frac{1}{2} ft^2$.

(2) A volume v of a gas at pressure p is enclosed in a cylindrical vessel of radius r. It is allowed to expand so that its volume is trebled. Find the total work done.

(3) A cork 3 ins. long is drawn from a bottle by a force proportional to the area remaining in contact with the bottle. If the initial pull $= 48$ lbs. weight, find the work done in extracting the cork.

(4) Extend the formula for work done in stretching an elastic string to cover the case in which the string is stretched from a length a (not its natural length) to a length b.

(5) A string 1 foot long is such that if one pound weight be suspended from it an extension of one inch is produced. Find the work done in stretching the string to 15 inches length.

(6) In the previous question the 1 lb. weight is allowed to fall momentarily through 1 foot before the string is taut; to what length will the string be stretched momentarily?

(7) A semi-circle is immersed vertically with its diameter in the free surface. Find the centre of pressure.

(8) A cubical box of side a has a hinged lid. The box is filled with water and tilted about the edge opposite to the hinge. Explain how to find the maximum angle through which we can tilt the cube without the water spilling.

CHAPTER X

DIFFERENTIAL EQUATIONS. A FEW TYPES

61. Differential Equations. General principles.

Equations containing $\dfrac{dy}{dx}$, $\dfrac{d^2y}{dx^2}$, ... and functions of y and x are known as Differential Equations. Such equations are of very frequent occurrence in mechanical and physical problems and a large branch of Higher Mathematics concerns itself with them. The determination of the nature of the relation between y and x which allows a given differential equation to be true is known as "solving the equation" or "finding the integral" of the equation. Space prohibits a long discussion of these equations, for further details of which the reader is referred to the standard text-books*.

One general principle becomes evident. It is clear that if the equation contains $\dfrac{d^n y}{dx^n}$, the solution must be equivalent to the performance, actual or concealed, of n successive integrations. Since each of these integrations introduces an arbitrary constant, the general solution of a differential equation of the nth order will contain n arbitrary constants.

Or the matter may be seen thus. Suppose the equation $f(x, y) = 0$ contains n undetermined constants. We can then differentiate n times, getting the n further equations

$$f_1(x, y) = 0, \quad f_2(x, y) = 0 \text{ and so on,}$$

* And in particular to Prof. H. T. Piaggio's excellent *Elementary Treatise on Differential Equations*.

each containing some or all of the arbitrary constants. We now have in all $(n + 1)$ equations containing these n constants and therefore we can in theory, at any rate, eliminate them. The eliminant will be an equation containing functions of x, y and derivatives of y up to and including the nth. That is, it will be a differential equation of the nth order; and so, conversely, we may expect that the solution of such an equation will lead us back to a relation containing n different arbitrary constants.

62. Types of solution already considered.

The process of integration is itself, of course, the same as finding the solution of an equation $\dfrac{dy}{dx} = f(x)$, and consequently any equation in which $\dfrac{dy}{dx}$ may be found in terms of x only will only need to be integrated directly, e.g. the equation

$$\left(\frac{dy}{dx}\right)^2 - x\frac{dy}{dx} + 6x^2 = 0$$

gives at once

$$\frac{dy}{dx} = 3x \text{ or } -2x,$$

$$\therefore \ y = \text{either } \tfrac{3}{2}x^2 + A \text{ or } -x^2 + B.$$

Similarly the equation $\dfrac{dy}{dx} = f(y)$ can be solved at once by inverting both sides, e.g.

$$\frac{dy}{dx} = \frac{y^2}{1 + y},$$

$$\therefore \ \frac{dx}{dy} = \frac{1 + y}{y^2} = \frac{1}{y^2} + \frac{1}{y},$$

$$\therefore \ x = -\frac{1}{y} + \log y + \text{const.}$$

And in general any equation in which the variables can be separated can be integrated as in the following example:

$$\frac{dy}{dx} = \frac{\sqrt{1-y^2}}{1+x^2},$$

$$\therefore \frac{dy}{\sqrt{1-y^2}} = \frac{dx}{1+x^2}.$$

Integrating both sides

$$\sin^{-1} y + C = \tan^{-1} x.$$

63. Linear Equations with Constant Coefficients.

We pass on to the consideration of a type which frequently occurs in Mechanical and Physical Problems.

The typical equation is

$$p_0 \frac{d^n y}{dx^n} + p_1 \frac{d^{n-1} y}{dx^{n-1}} + \ldots + p_n y = 0.$$

All the coefficients $p_0 \ldots p_n$ are constants, and no derivative appears except in the first power.

The simplest possible equation of this type is the equation

$$\frac{dy}{dx} = ky,$$

i.e.

$$\frac{dx}{dy} = \frac{1}{ky},$$

i.e.

$$x = \frac{1}{k} \log y + C.$$

For convenience let the arbitrary constant be written $\dfrac{-\log A}{k}$,

$$\therefore x = \frac{1}{k} \log \frac{y}{A},$$

$$\therefore \mathbf{y = Ae^{kx}}.$$

Notice that, as we expect, the solution contains one arbitrary constant.

64. Graphical representation of this result.

The reader should notice the two types of curve which can be represented by this equation according as k is positive or negative. If k is positive we have a curve in which y is always increasing at a rate proportional to its value at the instant considered. This type of relation is sometimes called "True Compound Interest." The graph is as in Fig. 22, the upper curve for a positive and the lower for a negative value of A, and

$$y \to \pm \infty \text{ as } x \to \infty,$$
$$y \to 0 \quad \text{as } x \to -\infty.$$

If k is negative the conditions are reversed, giving the two curves of Fig. 23. These are sometimes called "die-away curves."

The reader will do well to consider for himself the various types represented by $y = Ae^{kx} + B$.

65. Equations of the Second Order. The Auxiliary Equation.

We next consider equations of the form

$$p \frac{d^2y}{dx^2} + q \frac{dy}{dx} + ry = 0.$$

We shall expect that this type of equation will occur in Mechanics, for forces and accelerations are expressible in terms of second differentials.

The results found in the previous section suggest that exponential functions will form the basis of the solution.

Let us try as a solution $y = Ae^{\alpha x}$.

Substituting we get

$$Ae^{\alpha x} \{p\alpha^2 + q\alpha + r\} = 0.$$

Hence Ae^{ax} will be a solution when α is either of the roots of the "auxiliary" equation

$$p\alpha^2 + q\alpha + r = 0.$$

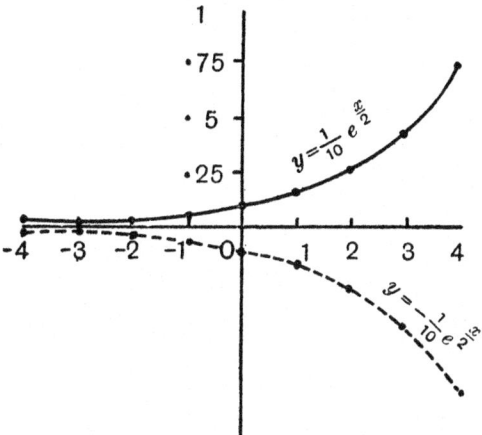

Form $y = Ae^{kx}$. k positive.

Fig. 22.

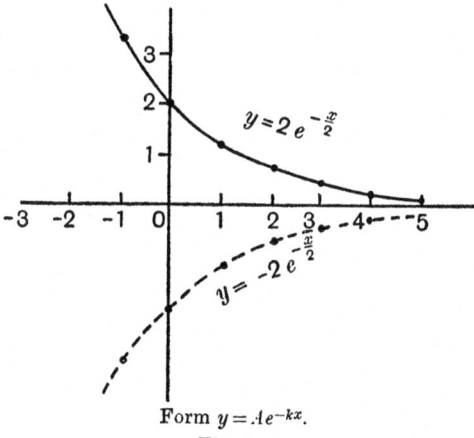

Form $y = Ae^{-kx}$.

Fig. 23.

Denoting these roots by α and β, $Ae^{\alpha x}$ and $Be^{\beta x}$ are both solutions and hence $Ae^{\alpha x} + Be^{\beta x}$ is seen to be a solution containing two arbitrary constants, i.e. a general solution.

The principle can clearly be extended to equations of any order.

$Ex.$ $$\frac{d^2y}{dx^2} + 7\frac{dy}{dx} + 10y = 0.$$

Auxiliary equation

$$\alpha^2 + 7\alpha + 10 = 0, \quad \text{roots} -5, -2.$$

$\therefore$ Solution is $\quad y = Ae^{-5x} + Be^{-2x}$.

66. Exceptional cases. Imaginary Roots.

It very often happens that the roots of the auxiliary equation are imaginary (whenever $q^2 < 4r$).

The general solution now reduces to the form

$$y = Ae^{(\alpha+\sqrt{-1}\beta)x} + Be^{(\alpha-\sqrt{-1}\beta)x}$$

or $$= e^{\alpha x}\{Ae^{\sqrt{-1}\beta x} + Be^{-\sqrt{-1}\beta x}\}.$$

But by Euler's Exponential Values,

$$\left. \begin{array}{l} 2\cos\beta x = e^{\sqrt{-1}\beta x} + e^{-\sqrt{-1}\beta x} \\ 2\sqrt{-1}\sin\beta x = e^{\sqrt{-1}\beta x} + e^{-\sqrt{-1}\beta x} \end{array} \right\} *.$$

Thus any expression $Ae^{\sqrt{-1}\beta x} + Be^{-\sqrt{-1}\beta x}$ can also be written in the form $A_1\cos\beta x + B_1\sin\beta x$, where A_1 and B_1 are two new arbitrary constants. The general solution in the case of imaginary roots is therefore expressed in the form

$$y = e^{\alpha x}\{A\cos\beta x + B\sin\beta x\}.$$

The case most frequently met with is that where the middle term of the equation is missing, leaving only the terms

$$\frac{d^2y}{dx^2} + ky = 0 \quad (k \text{ positive}).$$

* Hobson's *Trigonometry*, §§ 228, 229.

The roots being $\pm \sqrt{-k}$, the solution is

$$y = A \cos \sqrt{k}x + B \sin \sqrt{k}x,$$

a form which should be noted for future reference.

67. Equal Roots.

It may also occur that the auxiliary equation has equal roots; and the form of the general solution reduces to $(A + B)\,e^{ax}$, which only includes one true arbitrary constant and thus is not the general solution.

Consider the equation

$$\frac{d^2y}{dx^2} - 2a\,\frac{dy}{dx} + a^2y = 0.$$

The auxiliary equation has equal roots a, and the solution clearly involves e^{ax}.

Put $\qquad\qquad y = e^{ax} \cdot v.$

Substituting, the equation becomes

$$e^{ax} \cdot \frac{d^2v}{dx^2} = 0,$$

$$\therefore \frac{d^2v}{dx^2} = 0,$$

and $\qquad\qquad v = Ax + B.$

Hence the general solution is

$$y = e^{ax}\,(Ax + B).$$

It is clear that this method can be extended to equations of any order.

Ex. (i). $\qquad\qquad \dfrac{d^3y}{dx^3} + y = 0,$

roots of $a^3 + 1$ are $-1, -\dfrac{1}{2} \pm \dfrac{\sqrt{-3}}{2}$.

The solution is

$$y = Ae^{-x} + e^{-\frac{x}{2}}\left(B \cos \frac{\sqrt{3}\,x}{2} + C \sin \frac{\sqrt{3}\,x}{2} \right).$$

Ex. (ii).
$$\frac{d^2y}{dx^2} + 6\frac{dy}{dx} + 9y = 0.$$
$$(\alpha + 3)^2 = 0.$$
$$y = e^{-3x}(Ax + B).$$

68. Illustrations of preceding equations. Simple Harmonic Motion.

Simple Harmonic Motion is defined as the motion performed by a body, moving in a straight line under a force which is directed towards a fixed point in that line and proportional to the distance from it.

Taking the fixed point in the line as origin, the equation giving the distance of the point from the origin at time t is evidently $\dfrac{d^2x}{dt^2} = -kx$, the negative sign being prefixed since the acceleration is inward, i.e. opposes the increase of x.

We have seen that the general solution is
$$x = A\cos\sqrt{k}\,t + B\sin\sqrt{k}\,t.$$
Using a transformation given in Ch. III we may write this
$$x = (A^2 + B^2)^{\frac{1}{2}}\left\{\cos\left[\sqrt{k}\,t - \tan^{-1}\frac{B}{A}\right]\right\},$$

x has therefore a maximum value $\sqrt{A^2 + B^2}$ and a minimum value $-\sqrt{A^2 + B^2}$ and the whole motion takes place between these points. The quantity $\sqrt{A^2 + B^2}$ is called the Amplitude of the motion. Further, we can see that the values of x are repeated periodically, for
$$\cos\{\sqrt{k}\,t - \text{const.}\}$$
repeats its values when t has increased by $\dfrac{2\pi}{\sqrt{k}}$. The motion is therefore periodic and its period is $\dfrac{2\pi}{\sqrt{k}}$.

Initial conditions are usually given to determine A and B. For instance in the usual case of s.h.m. the moving point starts with zero velocity at zero time at $x = a$.

We thus get
$$\left. \begin{array}{l} x = a \text{ when } t = 0 \\ \dfrac{dx}{dt} = 0 \text{ when } t = 0 \end{array} \right\},$$

$$\therefore A = a, \quad B = 0,$$

and the equation reduces to the simple form $x = a \cos \sqrt{k}\, t$. This is clearly the motion performed by the foot of the perpendicular from a point on a circle to the x-axis, when the circle is described with constant angular velocity $\sqrt{k}$. This is the standpoint from which s.h.m. was first considered, and the student will find this treatment of the subject in elementary works on Dynamics.

69. General Oscillatory Motion. The Pendulum.

We notice, in fact, that any equation of the form $\dfrac{d^2\phi}{dt^2} = -\lambda\phi$, where ϕ is some parameter which fixes the position of the body at time t, represents a motion, the phases of which repeat themselves after an interval $\dfrac{2\pi}{\sqrt{\lambda}}$, or, as we say, represents an **Oscillation of period** $\dfrac{2\pi}{\sqrt{\lambda}}$.

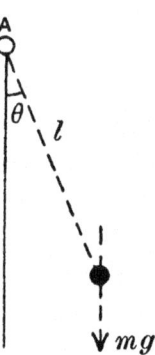

Fig. 24.

Consider, for example, the case of a pendulum formed by a small bob of mass m, at the end of a string of length l, oscillating through a small angle.

The position of the bob is thus determined entirely by the angle θ.

8

Writing down the equation of motion, as in the preceding chapter,

$$I \frac{d^2\theta}{dt^2} = L;$$

i.e. we have $$ml^2 \frac{d^2\theta}{dt^2} = - mgl \sin \theta,$$

i.e. $$\frac{d^2\theta}{dt^2} = - \frac{g}{l} \theta,$$

since θ is a small angle.

Therefore we thus see at once that the motion is periodic and its total period $2\pi \sqrt{\dfrac{l}{g}}$. We usually put this result in the form

Time of Vibration (half oscillation) $= \pi \sqrt{\dfrac{l}{g}}.$

The student of Physics will be familiar with many experiments in which a formula of this type is used to determine g*.

70. Resisted Motion. Damped Oscillation.

As a further example consider the following.

A body moves in a straight line under a central force proportional to the distance, but its motion is resisted by a force which is always proportional to the velocity. Determine the motion.

The equation will clearly be

$$\frac{d^2x}{dt^2} = - 2\lambda \frac{dx}{dt} - \mu x$$

(negative signs since both forces are retardations), i.e.

$$\frac{d^2x}{dt^2} + 2\lambda \frac{dx}{dt} + \mu x = 0.$$

* These formulæ are usually slightly more complicated, for I is usually the M.I. of a compound body, such as a sphere suspended by a string. Again the principle is more important than the formula.

Solving the auxiliary equation $\alpha^2 + 2\lambda\alpha + \mu = 0$ we get as the general solution

$$x = Ae^{(-\lambda+\sqrt{\lambda^2-\mu})t} + Be^{(-\lambda-\sqrt{\lambda^2-\mu})t}$$
$$= e^{-\lambda t}(Ae^{\sqrt{\lambda^2-\mu}\,t} + Be^{-\sqrt{\lambda^2-\mu}\,t}).$$

If, as is usually the case, $\lambda^2 < \mu$ (the resistance-constant usually being small), the radical $\sqrt{\lambda^2 - \mu}$ is imaginary, and following a previous article we have to write (§ 66)

$$x = e^{-\lambda t}(A\cos\sqrt{\mu-\lambda^2}.t + B\sin\sqrt{\mu-\lambda^2}.t)$$
$$= \sqrt{A^2+B^2}\,e^{-\lambda t}\left[\cos\left(\sqrt{\mu-\lambda^2}.t - \tan^{-1}\frac{B}{A}\right)\right].$$

Clearly both x and $\dfrac{dx}{dt}$ contain periodic functions, and maximum and minimum values, both of displacement and velocity, occur at intervals of $\dfrac{2\pi}{\sqrt{\mu-\lambda^2}}$.

The motion is semi-oscillatory in character but it possesses the continually decreasing amplitude $\sqrt{A^2+B^2}\,e^{-\lambda t}$.

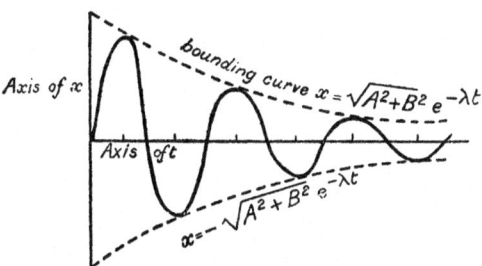

Fig. 25. $x = Ae^{-\lambda t}\cos\mu t$.

Such a motion is known as a Damped Oscillation, and occurs for instance in connection with the oscillatory discharges of Spark Wireless. It can be easily understood diagrammatically as in Fig. 25.

If however $\lambda^2 > \mu$, the solution will not contain trigono-metrical functions, but exponentials, and the motion is not oscillatory at all.

71. Equations in which the right-hand side is not zero.

The discussion of the equations in which the right-hand side is not zero, but a function of x, lies beyond the scope of this work. We can however briefly mention one general principle which will often provide the solution of such equations.

Consider the equation

$$\frac{d^2y}{dx^2} + 4\frac{dy}{dx} + 3y = 12 + 9x.$$

We see by inspection that $y = 3x$ is one solution.
Let $y = 3x + v$. Then by substitution

$$\frac{d^2v}{dx^2} + 4\frac{dv}{dx} + 3v = 0,$$

or v is a solution of the same equation with zero on the right-hand side.

We see thus that the solution of any such equation consists of a particular integral together with the Complementary Function (i.e. the general solution formed by writing zero on the right-hand side).

The complete solution of the example given is thus

$$\mathbf{y = Ae^{-3x} + Be^{-x} + 3x.}$$

The determination of these particular integrals often demands great ingenuity, and for various methods of doing this the reader is referred to any standard work on Differential Equations.

72. The Equation of Resonance.

We proceed to solve, however, one equation of this type which illustrates an important physical principle.

Consider the equation

$$\frac{d^2x}{dt^2} = -\mu x + L \cos pt.$$

(The equation represents the state of a body acted on by (1) an ordinary central force, and (2) a fixed periodic force.)

We suspect that $A \cos pt$ is a particular solution for some value of A and substituting we get $A = \dfrac{L}{\mu - p^2}$.

Thus $\dfrac{L \cos pt}{\mu - p^2}$ is a particular integral.

The solution is now

$$x = A \cos \sqrt{\mu}\, t + B \sin \sqrt{\mu}\, t + \frac{L}{\mu - p^2} \cos pt.$$

The motion is composed of two Simple Harmonic Motions, one of period $\dfrac{2\pi}{\sqrt{\mu}}$, the other of period $\dfrac{2\pi}{p}$. But it is clear that when μ and p^2 are nearly equal the last member $\dfrac{L}{\mu - p^2} \cos pt$ can become very large even though the actual value of L is small.

It is thus made clear that the superposition of two Simple Harmonic Motions (or two periodic oscillations) may produce an effect far out of proportion to the actual amplitude of either of the motions taken separately.

Thus soldiers have to break step crossing a bridge, in case the rhythm of their march should be equi-periodic with the natural vibration of the bridge; and the combination of the flute with the human voice sometimes produces a very resonant effect. The principle will also be familiar to the students of Wireless Telegraphy and Sound (Heterodyne Reception and Beats).

There are many special devices used in the solution of particular equations. One of these devices (that of multiplying both sides by dy/dx) has been used in a dynamical example of the preceding chapter and should be referred to again, as it is of use in many cases. (Cf. p. 98.)

EXAMPLES X

Solve the equations:

(1) $\dfrac{dy}{dx} + 4y = 0.$

(2) $\dfrac{dy}{dx} = 4y.$

(3) $\dfrac{dy}{dx} = 4y + 4.$

(4) $\dfrac{dy}{dx} = 4y + a.$ (a constant.)

(5) $\dfrac{d^2y}{dx^2} + 4\dfrac{dy}{dx} = 0.$

(6) $\dfrac{d^2y}{dx^2} + 4\dfrac{dy}{dx} = a.$

(7) The outward velocity of a particle moving in a line is 4 times its distance from a point in that line, diminished by 6. If the particle starts from unit distance at zero time, solve the equation of motion and represent the motion graphically.

(8) Solve the equation given by conditions similar to (7) with "inward" velocity substituted for "outward."

Solve the equations:

(9) $\dfrac{d^2y}{dx^2} + 5\dfrac{dy}{dx} + 6y = 0.$

(10) $\dfrac{d^2y}{dx^2} + 5\dfrac{dy}{dx} + 6y = 12.$

(11) $\dfrac{d^2y}{dx^2} - 6\dfrac{dy}{dx} + 9y = 0.$

(12) $\dfrac{d^2y}{dx^2} - 6\dfrac{dy}{dx} + 9y = e^x.$

(13) $\dfrac{d^2y}{dx^2} + 4y = 0.$

(14) $\dfrac{d^2y}{dx^2} - 4\dfrac{dy}{dx} + y = 0.$

(15) $\dfrac{d^3y}{dx^3} + 6\dfrac{d^2y}{dx^2} + 11\dfrac{dy}{dx} + 6y = 0.$

(16) $\dfrac{d^3y}{dx^3} - 3\dfrac{dy}{dx} + 2y = 4x - 6.$

(17) $\dfrac{d^4y}{dx^4} - 16y = 0.$

(18) An elastic string (mod λ) of length l has a bob of mass m attached to the end. Show that if the bob be set in motion by stretching the string it will perform a Simple Harmonic Motion, and find the period.

(19) Explain how the velocity of a bullet may be found by firing it into the bob of a known pendulum.

(20) Find the period of a pendulum composed of a heavy rod of length $2a$ suspended from one end.

(21) Find the period of the pendulum formed by a heavy sphere of radius r attached to a weightless string of length l.

(22) Integrate the equation $\dfrac{d^2s}{dt^2} = f$ to get the usual formulæ of accelerated motion.

(23) Show that for any law of air-resistance of the form Av^n (v is velocity at time t) a falling body will have a maximum or terminal velocity.

(24) Integrate the equation of motion of a falling body when the resistance varies as the velocity.

Find the time taken to reach the terminal velocity if the body falls from rest.

ANSWERS TO EXAMPLES

Examples marked with an asterisk are taken from Higher Certificate Papers by permission of the Board.

(Arbitrary constants to be understood in all cases.)

I A

(1) $\frac{2}{3}x^{\frac{3}{2}}$; $2x^{\frac{1}{2}}$; $\frac{x}{3}+\frac{x^2}{6}+\frac{x^3}{9}$.　　(2) $\frac{3}{7}x^{\frac{7}{3}}$; $-\frac{1}{5}x^{-5}$; $-\frac{5}{2}x^{-\frac{2}{5}}$.

(3) $\frac{1}{4}(3x-4)^{\frac{4}{3}}$; $\frac{2}{9}(3x-4)^{\frac{3}{2}}$; $\frac{2}{3}(3x-4)^{\frac{1}{2}}$.

(4) $\log(a+x)$; $\log\dfrac{a+x}{a-x}$; $\dfrac{1}{(n-1)(a-x)^{n-1}}$.

(5) $\log(x+1)$; $x+2\log(x+1)$; $\dfrac{1}{a}\log(ax+b)$;

$\quad x+\dfrac{c-b}{a}\log(ax+b)$; $\dfrac{A}{a}x+\dfrac{Ba-bA}{a^2}\log(ax+b)$.

(6) $2x^{\frac{1}{2}}+\frac{2}{3}x^{\frac{3}{2}}$; $\frac{1}{2}x^2+\log x$; $\frac{2}{3}\left[(x+1)^{\frac{3}{2}}-x^{\frac{3}{2}}\right]$; $\frac{2}{5}(x+1)^{\frac{5}{2}}-\frac{2}{3}(x+1)^{\frac{3}{2}}$.

(8) $\log(x+\sqrt{x^2-a^2})$.

I B

(1) $\sin^2 x$; $\dfrac{x}{2}+\dfrac{\sin 2x}{4}$; $\dfrac{1}{2}\left[\dfrac{\sin(m+n)x}{m+n}+\dfrac{\sin(m-n)x}{m-n}\right]$;

$\quad \dfrac{1}{2}\left[\dfrac{\sin(m-n)x}{m-n}-\dfrac{\sin(m+n)x}{m+n}\right]$.

(2) $\log\sec x$; $\sec x$; $\log\sin x-\frac{1}{2}\operatorname{cosec}^2 x$; $\log\tan x$.

(3) $\log\log x$; $\log(e^x+2)$; $\log(e^x-e^{-x})$; $\log(\log\sin x)$.

(4) $\log(x^2-3x+2)$; $x+\log(2x-1)$; $\frac{1}{4}(x^2+2x+2)^2$;

$\quad \dfrac{x^2}{2}+x+\log(x+1)$; $\frac{1}{2}\log(x^2+2x+2)$.

(5) $\frac{1}{3}\tan^{-1}\dfrac{x}{3}$; $x-2\tan^{-1}\dfrac{x}{2}$; $\tan^{-1}x^2$; $\dfrac{1}{\sqrt{21}}\tan^{-1}\dfrac{x\sqrt{3}}{\sqrt{7}}$.

(6) $-\frac{1}{4}\cos^4 x$; $\frac{1}{3}\sin^3 x-\frac{1}{5}\sin^5 x$; $\sin x-\frac{1}{3}\sin^3 x$.

II

(1) $-\frac{2}{3}(1-x)^{\frac{3}{2}}$.　　　(2) $\frac{3}{8}(1+2x)^{\frac{4}{3}}$.　　　(3) $\tan^{-1}x^2$.

(4) $\tan^{-1}x^3$.　　　(5) $\frac{1}{\sqrt{2}}\tan^{-1}\dfrac{x+1}{\sqrt{2}}$.　　　(6) $\sin^{-1}\dfrac{x+1}{3}$.

(7) $\sec^{-1}(x-1)$.　　　(8) $\sin^{-1}(2x-1)$ $(z=x-\frac{1}{2})$.

(9) $\tan^{-1}e^x$ (get rid of negative powers).　　　(10) $\sin e^x$.

(11) a.　　　(12) $\frac{2}{9}a^{\frac{9}{2}}(2\sqrt{2}-1)$.　　　(16) $\tan\dfrac{x}{2}$.

(17) $-\dfrac{1}{a}\sqrt{\dfrac{2a-x}{x}}$.　$\left(\text{Put } y=\dfrac{1}{x}.\right)$

(18) $\dfrac{1}{a}\sec^{-1}\dfrac{x}{a}$.　　(19) $\sin^{-1}\dfrac{x-a}{a}$.　　(20) $\frac{1}{2}\tan^2 x$.　　(21) $\dfrac{\pi}{32}$.

III

(1) $-(x+1)e^{-x}$; $e^{2x}\left(\dfrac{x}{2}-\dfrac{1}{4}\right)$; $e^x(x^2-2x+2)$;

$e^x(x^3-3x^2+6x-6)$.

(2) $\sin x-x\cos x$; $(x^2-2)\sin x+2x\cos x$;

$(6x-x^3)\cos x+3(x^2-2)\sin x$; $\dfrac{x^2+x\sin 2x}{4}+\dfrac{\cos 2x}{8}$;

$\dfrac{\sin 2x}{8}-\dfrac{x\cos 2x}{4}$.

(3) $\dfrac{x^2}{2}\log x-\dfrac{x^2}{4}$; $\dfrac{x^3}{3}\log x-\dfrac{x^3}{9}$; $\dfrac{x^{n+1}}{n+1}\log x-\dfrac{x^{n+1}}{(n+1)^2}$.

(4) $\frac{1}{2}e^x(\sin x+\cos x)$; $\dfrac{1}{\sqrt{5}}e^x\cos(2x-\tan^{-1}2)$;

$\dfrac{1}{\sqrt{5}}e^x\sin(2x-\tan^{-1}2)$; $e^x\sin x$.

(5) $x\sin^{-1}x+\sqrt{1-x^2}$; $x\tan^{-1}x-\frac{1}{2}\log(1+x^2)$;

$\dfrac{x^2}{2}\sin^{-1}x+\frac{1}{4}x\sqrt{1-x^2}-\frac{1}{4}\sin^{-1}x$; $\dfrac{x^2+1}{2}\tan^{-1}x-\dfrac{x}{2}$.

(6) $\dfrac{x\sqrt{x^2-a^2}}{2}-\dfrac{a^2}{2}\log\dfrac{x+\sqrt{x^2-a^2}}{a}$; $\dfrac{x\sqrt{a^2-x^2}}{2}+\dfrac{a^2}{2}\sin^{-1}\dfrac{x}{a}$.

(7) $\dfrac{e^{3x}}{5}\left\{\cos\left(4x-\tan^{-1}\tfrac{4}{3}\right)\right\}$;

$e^{5x}\left\{\dfrac{1}{5\sqrt{5}}\sin\left(10x-\tan^{-1}2\right)+\dfrac{1}{\sqrt{41}}\sin\left(4x-\tan^{-1}\tfrac{4}{5}\right)\right\}$;

$e^{x}\left\{\dfrac{1}{\sqrt{5}}\sin\left(2x-\tan^{-1}2\right)+\dfrac{1}{\sqrt{17}}\sin\left(4x-\tan^{-1}4\right)\right.$

$\left.\qquad\qquad\qquad-\dfrac{1}{\sqrt{37}}\sin\left(6x-\tan^{-1}6\right)\right\}$.

(8) $x\tan x+\log\cos x$; $\dfrac{x}{\sqrt{1-x^2}}\sin^{-1}x+\log\sqrt{1-x^2}$.

(Put $y=\sin^{-1}x$ in last example.)

IV

(1) $\frac{1}{2}\log\dfrac{x+1}{x+3}$. 　　　　(2) $\frac{1}{2}\log(x^2+2x+3)-\dfrac{1}{\sqrt{2}}\tan^{-1}\dfrac{x+1}{\sqrt{2}}$.

(3) $\dfrac{1}{25}\left[\log(4-5x)+\dfrac{4}{4-5x}\right]$. 　　(4) $\log\dfrac{(x-3)^3}{(x-2)^2}$.

(5) $2x-\frac{9}{2}\log(x^2+6x+10)+11\tan^{-1}(x+3)$.

(6) $x-2\log(x^2+2x+2)+3\tan^{-1}(x+1)$.

(7) $\frac{1}{2}\log(x^2-2x-2)-\log(x-1)$.

(8) $\frac{2}{3}\log(x+1)+\frac{1}{6}\log(x^2-x+1)+\dfrac{1}{\sqrt{3}}\tan^{-1}\dfrac{2x-1}{\sqrt{3}}$.

(9) $\log x-\frac{1}{2}\log(x^2+x+1)-\dfrac{1}{\sqrt{3}}\tan^{-1}\dfrac{2x+1}{\sqrt{3}}$. 　　(10) $1+\dfrac{2\pi}{3\sqrt{3}}$.

(11) $\dfrac{1}{7}\left(\log\dfrac{x-2}{x+2}-\sqrt{3}\tan^{-1}\dfrac{x}{\sqrt{3}}\right)$.

(12) $\dfrac{1}{ab(a^2-b^2)}\left(a\tan^{-1}\dfrac{x}{b}-b\tan^{-1}\dfrac{x}{a}\right)$.

(13) $\dfrac{1}{4\sqrt{2}}\log\dfrac{x^2+\sqrt{2}.x+1}{x^2-\sqrt{2}x+1}+\dfrac{1}{2\sqrt{2}}\tan^{-1}\dfrac{x\sqrt{2}}{1-x^2}$.

(14) $\frac{1}{24}\log\dfrac{4+3\sin x}{4-3\sin x}$.

(15) $\frac{1}{4}\log_e\frac{5}{3}$. 　　　　　　(16) $\dfrac{\pi}{2(a+b)}$.

V A

(1) $\frac{1}{3}\cos^3 x - \cos x.$

(2) $-\frac{1}{5}\cos^5 x + \frac{2}{3}\cos^3 x - \cos x.$

(3) $\frac{1}{9}\sin^9 x - \frac{4}{7}\sin^7 x + \frac{6}{5}\sin^5 x - \frac{4}{3}\sin^3 x + \sin x.$

(4) $\frac{1}{3}\sin^3 x - \frac{2}{5}\sin^5 x + \frac{1}{7}\sin^7 x.$

(5) $\sin x + 2\operatorname{cosec} x - \frac{1}{3}\operatorname{cosec}^3 x.$

(6) $\dfrac{1}{4}\log\dfrac{1+\cos x}{1-\cos x} - \dfrac{1}{2}\dfrac{\cos x}{\sin^2 x}.$

(7) $-\frac{1}{3}\cos^6 x.$

(8) $\frac{1}{5}\cos^5 x - \frac{4}{7}\cos^7 x.$

(9) $\dfrac{1}{4}\log\dfrac{2+\tan\frac{x}{2}}{2-\tan\frac{x}{2}}.$

(10) $\dfrac{1}{4}\log\dfrac{3-\tan\frac{x}{2}}{1-3\tan\frac{x}{2}}.$

(11) $\dfrac{1}{\sqrt{2}}\log\dfrac{1-\sqrt{2}+\tan\frac{x}{2}}{1+\sqrt{2}-\tan\frac{x}{2}}.$

(12) $\log\left(1+\tan\frac{x}{2}\right).$

(13) $\sqrt{2}\tan^{-1}\dfrac{1+\tan\frac{x}{2}}{\sqrt{2}}.$

(14) $\dfrac{1}{\sqrt{a^2+b^2}}\log\dfrac{a\tan\frac{x}{2}-b+\sqrt{a^2+b^2}}{a\tan\frac{x}{2}-b-\sqrt{a^2+b^2}}.$

(17) $\dfrac{\pi-a}{\sin a}.$

(18) (a) $\dfrac{\pi}{2\sqrt{1-k^2}}$; $k<1.$ (b) $\dfrac{1}{2\sqrt{k^2-1}}\log\left(1-\dfrac{1}{k^2}\right)$; $k>1.$

(19) $\dfrac{1}{\sin a}\log\dfrac{1+\tan\frac{x}{2}\tan\frac{a}{2}}{1-\tan\frac{x}{2}\tan\frac{a}{2}}.$

V B

(1) $\dfrac{3x}{8} - \dfrac{\sin 2x}{4} + \dfrac{\sin 4x}{32}.$

(2) $\dfrac{5x}{16} + \frac{15}{64}\sin 2x + \frac{3}{64}\sin 4x + \frac{1}{192}\sin 6x.$

(3) $\dfrac{x}{8} - \dfrac{\sin 4x}{32}$.   (4) $\dfrac{x}{16} - \dfrac{\sin 2x}{64} - \dfrac{\sin 4x}{64} + \dfrac{\sin 6x}{192}$.

(5) $\tan x - \cot x$.

(6) $-\frac{1}{2}(\cot x \operatorname{cosec} x) - \frac{1}{2}\log(\cot x + \operatorname{cosec} x)$.   (Put $c = \cot x$.)

(7) $\frac{1}{3}\sec^3 x - \frac{1}{2}\log\dfrac{1+\cos x}{1-\cos x}$.

(8) $I_n = \dfrac{1}{n+1}\sin^{n+1} x \cos x + \dfrac{n+2}{n+1} I_{n+2}$.

(9) (a) $I_{p,\,q} = \dfrac{1}{p+1}\sin^{p+1} x \cos^{q-1} x + \dfrac{q-1}{p+1} I_{p+2\ q-2}$.

(b) $I_{p,\,q} = \dfrac{1}{q+1}\sin^{p-1} x \cos^{q+1} x + \dfrac{p-1}{q+1} I_{p-2,\,q+2}$.

(10) $\dfrac{35\pi}{256}, \dfrac{35\pi}{128}, \dfrac{35\pi}{128}$.   (11) $\dfrac{5\pi}{32}, \dfrac{16}{35}, 0, 0$.

(12) $\dfrac{3\pi}{256}, \dfrac{3\pi}{128}$.   (13) 0.

(15) $\frac{1}{2}\tan\phi\sec\phi + \frac{1}{2}\log\tan\left(\dfrac{\pi}{4} + \dfrac{\phi}{2}\right)$;

$\frac{1}{4}\tan\phi\sec^3\phi + \frac{3}{8}\tan\phi\sec\phi + \frac{3}{8}\log\tan\left(\dfrac{\pi}{4} + \dfrac{\phi}{2}\right)$.

(16) $\dfrac{x^5(\log x)^2}{5} - \dfrac{2x^5(\log x)}{5^2} + \dfrac{2x^5}{5^3}$.

(17) $I = -\dfrac{1}{m+p+1}\cdot x^{m+1}(a^2-x^2)^{\frac{p}{2}}$

$+ \dfrac{pa}{m+p+1}\displaystyle\int x^m (a^2-x^2)^{\frac{p}{2}-1}\,dx.$

(19) $I_m = -\dfrac{x^{m+\frac{1}{2}}\sqrt{(2a-x)^3}}{m+2} + \dfrac{2m+1}{m+2}\cdot a I_{m-1}$.   (20) $\dfrac{21\pi a^6}{16}$.

VI

(1) (a) $11\frac{1}{2}$ units of area.   (b) $15\frac{1}{2}$ units of area.

(2) (a) $\frac{3}{4} a^{\frac{2}{3}} h^{\frac{4}{3}}$ (b) $\dfrac{h^4}{4a^2} + \dfrac{h^3}{3a}$.   (3) 2 units of area.   (4) $\dfrac{16a^2}{3}$.

(6) The integration of $\sin^2\theta$ on which it depends assumes the formulæ of Radian measure, which are equivalent to the formulæ for area and circumference of a circle.

(7) $\frac{8}{15}$ units of area.   (8) πa^2.   (10) $2:3$.   (11) $\frac{44}{15}$ units of area.

(12) $2a^2\left(1-\dfrac{\pi}{4}\right)$; $2a^2\left(1+\dfrac{\pi}{4}\right)$.   (13) $4a^2$.

VII A

(1) $\dfrac{\pi a^2}{16}$.   (2) $\dfrac{\pi\,(a^2+b^2)}{2}$.   (3) $\dfrac{a^2}{2}$.   (4) $\pi\left(a^2+\dfrac{b^2}{2}\right)$.

(5) (i) $\dfrac{3\pi a^2}{2}$; (ii) $\left(a^2+\dfrac{b^2}{2}\right)\left(\pi-\cos^{-1}\dfrac{a}{b}\right)$.   (6) $\dfrac{a^2}{4}(\epsilon^{2\pi}-1)$.

(7) $\frac{1}{3}\pi h\,(r_1{}^2+r_1 r_2+r_2{}^2)$.

(8) $\pi\left[a^2\{(a^2-r_1{}^2)^{\frac{1}{2}}-(a^2-r_2{}^2)^{\frac{1}{2}}\}-\frac{1}{3}\{(a^2-r_1{}^2)^{\frac{3}{2}}-(a^2-r_2{}^2)^{\frac{3}{2}}\}\right]$.

(9) $\frac{4}{3}\pi ab^2$.

(10) The spheroid is "prolate," i.e. of revolution about the minor axis; $\therefore$ its volume is $\frac{4}{3}\pi a^2 b$. Hence the error is $\frac{4}{3}\pi a^2(a-b)$ in excess, i.e. $\left(\dfrac{a-b}{b}\right)100\,°/_{\circ}=100\left(\dfrac{a}{b}-1\right)°/_{\circ}$. But $a^2-b^2=a^2\cdot\frac{1}{3600}$; $\therefore\dfrac{a}{b}=(1-\frac{1}{3600})^{-\frac{1}{2}}=1+\frac{1}{7200}$ approx. $\therefore$ error $=\frac{1}{72}\,°/_{\circ}$.

(11) $2\pi a h^2$.   (12) $\dfrac{\pi^2 a^3}{2}$.   (13) $\pi c^2\left(r-\dfrac{c}{3}\right)$.   (14) $\dfrac{128\pi r^3}{375}$.

(15) Where $x=\dfrac{144a}{49}$.

VII B

(2) The arc from $\theta=\beta$ to $\theta=a$ is

$$a\left[\frac{\theta\sqrt{\theta^2+1}}{2}+\tfrac{1}{2}\log\,(\theta+\sqrt{\theta^2+1})\right]_a^\beta.$$

(3) $4a$.   (4) $\dfrac{4b\,(a+b)}{a}\cos\dfrac{a}{2b}\,\theta_1$.   (5) $2a\log\dfrac{a+a}{a-a}-a$.

(6) $\dfrac{1}{27\sqrt{a}}\left[(4a+9x)^{\frac{3}{2}}\right]_{x_1}^{x_2}$.   (7) $6a$.

VIII

(1) At a distance $\dfrac{4r}{3\pi}$ along the bisecting radius.

(2) At a distance $\dfrac{2r}{\pi}$ from centre.

(3) Along the axis of the hemisphere at a distance $\dfrac{3r}{8}$.

(4) At $\left(\dfrac{4a}{3\pi}, \dfrac{4b}{3\pi}\right)$. 　　　　　(7) $\bar{x}=\dfrac{8a}{5m^2}$, $\bar{y}=\dfrac{2a}{m}$.

(8) Taking the planes at distances d_1 and d_2 from the centre the expression for $\bar{x}$ is

$$\frac{d_1+d_2}{2}\cdot\frac{a^2-\frac{1}{2}\,(d_1{}^2+d_2{}^2)}{a^2-\frac{1}{3}\,(d_1{}^2+d_1 d_2+d_2{}^2)}.$$

The surface $= \pi a\,(d_1-d_2)$.

(9) On the axis at a distance $\dfrac{h}{4}\cdot\dfrac{r^2+2rR+3R^2}{r^2+rR+R^2}$ from the smaller end.

(10) Surface $=\pi r\sqrt{h^2+r^2}$; c.g. $\frac{2}{3}h$ along axis.

(11) c.g. divides rod in ratio 2 : 1. 　　　(12) $\bar{x}=h\cdot\dfrac{m+3n}{2m+4n}$.

(13) $V=2\pi^2a^2d$; $S=4\pi^2ad$. 　　(14) $V=\pi a^3\sqrt{2}$; $S=4\pi a^2\sqrt{2}$.

(15) $\dfrac{\pi^2a^3}{12}$. 　　(17) $V=\cdot32625\pi$ cu. in. ; $S=4\cdot36\pi$ sq. ins. approx.

IX A

(1) $\dfrac{Ma^2}{3}$; $\dfrac{Mb^2}{3}$. 　　　　　(2) $M\dfrac{b^2+c^2}{3}$; $M\dfrac{c^2+a^2}{3}$; $M\dfrac{a^2+b^2}{3}$.

(3) $M\dfrac{r^2}{2}$. 　　(4) $\dfrac{2Ma^2}{3}$. 　　(5) $\dfrac{Mb^2}{4}$; $\dfrac{Ma^2}{4}$. 　　(6) $M\dfrac{r^2}{2}$.

(7) $\dfrac{5Ma^2}{4}$. 　　(8) $\dfrac{3Ma^2}{2}$. 　　(9) $M\dfrac{r^2}{3}$ ($2r=$base of $\triangle$).

(10) $\dfrac{2M}{5}\cdot\dfrac{b^5-a^5}{b^3-a^3}$. 　　(11) $\dfrac{5^7\pi^2}{36}$ ft. lbs. ; $\dfrac{5^6\pi}{288}$ secs. ; $\dfrac{5^7\pi}{864}$ revs.

(12) $\dfrac{5^6\pi}{576}$ secs.

(13) k.e. $=\dfrac{3M}{2}\left(\dfrac{d\theta}{dt}\right)^2$; p.e. $=-\dfrac{3M}{2}g\sin\theta$. Angular velocity $=\sqrt{g}$

radians per sec.

IX B

(1) Measure the area under the curve between ordinates corresponding to the given intervals.

(2) $pv \log 3$. 　　　　　　(3) 6 foot lbs.

(4) Extension × mean of initial and final tensions.

(5) $\frac{3}{8}$ foot lbs. 　　(6) To $\left(1 + \frac{1}{\sqrt{3}}\right)$ feet. 　　(7) $\frac{3}{8} r$.

(8) The angle is such that the moment about the hinge of the resultant pressure acting at the centre of pressure is equal to the moment of the weight of the lid.

X

(1) $y = A e^{-4t}$. 　　(2) $y = A e^{4t}$. 　　(3) $y = A e^{4t} + 1$.

(4) $y = A e^{4t} + \frac{a}{4}$. 　　(5) $y = A e^{-4t} + B$. 　　(6) $y = A e^{-4t} + B + \frac{ax}{4}$.

(7) $x = \frac{5}{2} e^{4t} + \frac{3}{2}$. 　　(8) $x = -\frac{1}{2} e^{-4t} - \frac{3}{2}$. 　　(9) $y = A e^{-3x} + B e^{-2x}$.

(10) $y = A e^{-3x} + B e^{-2x} + 2$. 　　(11) $y = e^{3x}(Ax + B)$.

(12) $y = e^{3x}(Ax + B) + \frac{e^x}{4}$. 　　(13) $y = A \cos 2x + B \sin 2x$.

(14) $y = e^{2x}(A \cos \sqrt{3}\, x + B \sin \sqrt{3}\, x)$.

(15) $y = A e^{-x} + B e^{-2x} + C e^{-3x}$. 　　(16) $y = e^x (Ax + B) + C e^{-2x} + 2x$.

(17) $y = A e^{2x} + B e^{-2x} + C \cos 2x + D \sin 2x$.

(18) Period $= 2\pi \sqrt{\dfrac{lm}{\lambda g}}$. 　　(20) $2\pi \sqrt{\dfrac{4a}{3g}}$.

(21) $2\pi \sqrt{\dfrac{l^2 + 2lr + \frac{7}{5} r^2}{(l + r) g}}$.

(23) For the downward acceleration $= g - \dfrac{A v^n}{m}$, which $= 0$ when $v = \sqrt[n]{\dfrac{mg}{A}}$, which will thus be the max. velocity.

(24) The equation is $\dfrac{d^2 x}{dt^2} = -\lambda \dfrac{dx}{dt} + g$ (x is the vertical height at time t) and the terminal velocity $\frac{g}{\lambda}$ is only reached after infinite time.